BRIAN McGOLDRICK

UNDERSTANDING ENVIRONMENTAL ISSUES

RESOURCES & POLLUTION

Hodder & Stoughton
LONDON SYDNEY AUCKLAND TORONTO

Acknowledgements

The author and publishers thank the following for permission to reproduce photographs and material in this book:

Barnabys Picture Library 3.1, 13.4, 33.2, 35.1
NCB 3.2
Author 4.1, 4.2, 4.4
Ecoscene 7.1, 8.2, 26.2
J Allan Cash 11.5, 15.1 (2), 20.2, 20.3, 23.1, 27.1, 30.1, 35.1
Association of Danish Windfarm Manufacturers 11.1
Danish Energy Agency 11.4
National Power 12.1
Warmer Bulletin 13.2
Simon Fraser 15.1
Sundernes Regional Heating 15.3
Sunderland & South Shields Water Co 17.1
Greenpeace 21.2
Mark Boulton/ICCE 22.1
National Farmers Union/Massey Ferguson 24.1
Dr R Evans 24.1
R R Furness 24.3
David K Jones 24.5, 24.7, 35.1
G W Moss 26.1
Fred Mayer/Magnum Picture Library 29.3
IDG 30.2, 30.4, 30.5
Popperfoto 31.1
NASA 32.1, 32.4

Every effort has been made to contact the holders of copyright material but if any have been inadvertently overlooked the publisher will be pleased to make the necessary alterations at the first opportunity.

British Library Cataloguing in Publication Data

McGoldrick, Brian
    Resources and pollution. — (Understanding
    environmental issues)
    I. Title II. Series
    333.7

    ISBN 0-340-52784-6

First published 1991

Typeset by Taurus Graphics, Abingdon, Oxon.
Printed in Hong Kong for the educational publishing division of Hodder and Stoughton Ltd, Mill Road, Dunton Green, Sevenoaks, Kent by Colorcraft.

Preface

Contents

To the teacher

Environmental education is an important part of the National Curriculum and will be delivered mainly by Geography aided by Science.

The majority of environmental examples identified in the Geography Attainment Targets can usefully be grouped under two general headings

* **Resources and Pollution**
* **Cities and Countryside.**

This book aims to address those environmental issues grouped under Resources and Pollution. Many of these issues are also identified in Science ATs 5 and 13 which respectively deal with Human Influences and Energy.

A second book, **Cities and Countryside**, will deal with the remaining environmental issues. Together, both books cover most environmental issues identified in the National Curriculum.

Using this book

Core questions

Generally, these aim to transmit knowledge and understanding, assist with skills, and encourage an interest in and care for the environment.

A and B questions

These mainly aim to provide practice in answering GCSE type questions.
B questions are more demanding.

Teachers should be selective in choosing questions for their classes.

Brian McGoldrick.

Introduction

unit 1

The 1800s

They did not know

Most people living in Britain last century did not fully understand that many of their activities caused harm to the environment. They did not know . . . so would you forgive them?

The 1900s

They did not care

For much of this century many people did know that some of their activities caused damage to the environment. Often they did little or nothing about it. They did know . . . would you forgive them?

A future history lesson

What about you?

Recently, there has been a great upsurge in environmental awareness. Many environmental issues have been brought to the attention of the public. By now you should know . . . **but do you care?**

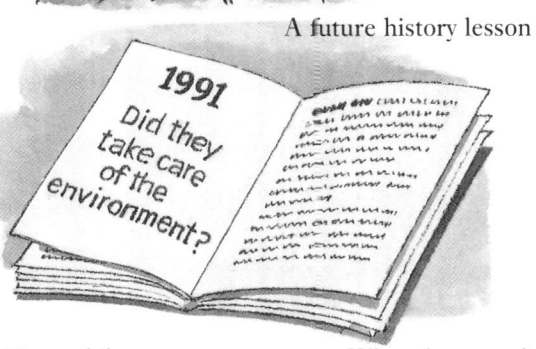

When you are history

In another 100 years you will be part of history. It will be the turn of your generation to be judged. Will pupils in 100 years' time say of you 'They knew, they cared, but they still did nothing about it', or 'They knew, they cared, and they did something about it'?

Figure 1.1 What about you?

The Department of the Environment carried out a survey in 1989 to find out how much people were worried about 21 environmental issues. Similar pictures to those drawn in figure 1.2 were used to show people each issue. Table 1a gives the results of the survey.

Concern for the environment

Figure 1.2   Environmental issues

Issue	Very worried %	Quite worried %
Chemicals put in rivers and the sea	64	27
Sewage contamination of bathing beaches	59	30
Radioactive waste	58	22
Destruction of the ozone layer	56	27
Oil spills at sea and oil on beaches	53	33
Insecticides, fertilizers and chemical sprays	46	35
Loss of wildlife and habitats. Species destruction	45	37
Destruction of tropical rainforests	44	31
Warming of the atmosphere by 'Greenhouse Effect'	44	28
Quality of drinking water	41	30
Acid rain	40	35
Fumes and smoke from factories	34	38
Traffic exhaust fumes	33	42
Litter and rubbish	33	41
Loss of trees and hedgerows	33	38
Fouling by dogs	29	31
Losing 'green belt' land	27	40
Decay of inner cities	22	38
Vacant and derelict land and buildings	16	34
Lack of access to open spaces and countryside	15	34
Noise	13	26

Table 1a   Concern about environmental issues 1989

Q U E S T I O N S

1 Which three environmental issues were most people:-
(a) quite worried about?
(b) very worried about?

2 What percentage of people were:-
(a) quite worried about acid rain?
(b) very worried about oil pollution?

3 Carry out a class survey to find out which environmental issues worry your class.

4 Design a questionnaire to find out which environmental issues concern either your teachers or local people. Figure 1.2 may be of help to you.

5 In small groups collect newspaper cuttings of environmental issues under these headings– Energy, Water Pollution, Air Pollution, Wildlife and Countryside, Other Issues.
Use the cuttings to make class wallcharts.

Bad news ... good news

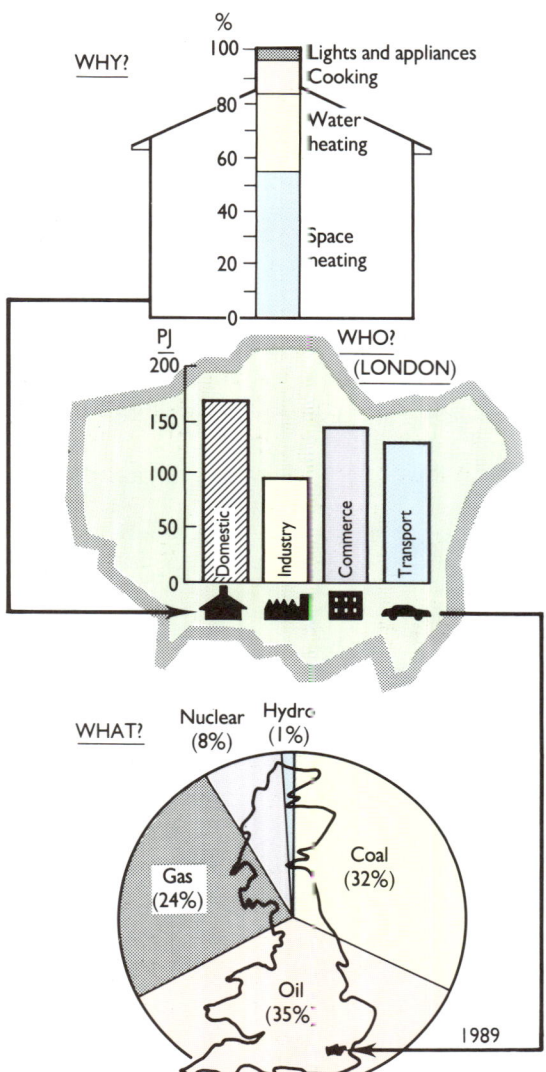

Figure 2.1   What? who? why?

	Years	
Finite fuel	UK	World
Oil	40–50	40
Gas	40–50	60
Coal	300	250

Table 2a   How long will they last?

How important is energy?

Energy is a very important resource. It:-

▶ Provides heat and light in homes, offices and shops.

▶ Fuels cars, lorries, trains and planes.

▶ Drives machines which make the things we eat and use.

▶ Powers televisions, kettles and cookers.

Can we do without it? Figure 2.1 shows:

▶ Why energy is used in one house in London.

▶ Who uses energy in all of London.

▶ What fuels are used in Britain.

Why is energy an important environmental issue?

Figure 2.2 shows the two main reasons.

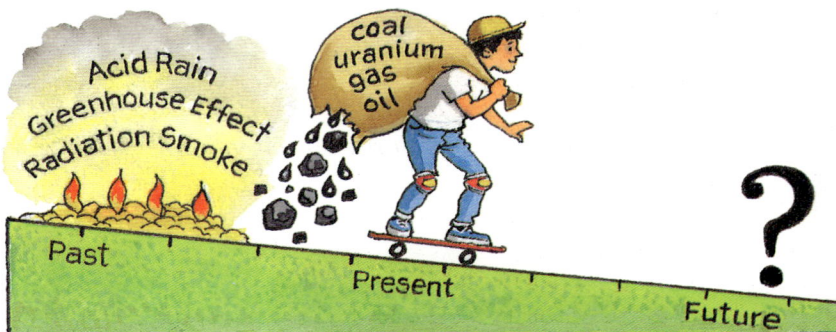

Figure 2.2   Running out

1 We are running out of some forms of energy.

2 Many of today's environmental problems are caused by the way we obtain and use energy.

3 Countries need secure energy supplies.

Bad news first

There is little doubt that we will run out of coal, oil and gas at some time in the future. These are called **finite** fuels because when they are burned they are finished, gone forever.

The Department of Energy has forecast that Britain will need 370 million tonnes of coal equivalent (mtce) of energy by 2025AD. Table 2b shows how much energy the Department thinks the renewables could supply by 2025AD.

Renewable energy source	Contribution by 2025AD (mtce)
Wave	0
Wind	1.6
Tidal	4.2
Hydro	1.0
Geothermal	0.3
Solar	2.1
Biofuel	3.6

Table 2b   Department of Energy forecast

Renewable energy		Technical	Resource	Economic	Electricity GWh/yr
Biofuels	Forestry	+	+	+	700
	Landfill Gas	+	+	+	300
	General Ind. Waste	+	+	+	2150
	Special Ind. Waste	+	+	+	60
	Municipal Waste	+	+	+	1900
Geothermal	Hot Dry Rocks	o	+	o	4500
	Aquifers	+	o	+	0
Water	Small Scale Hydro	+	+	+	55
	Tidal	+	+	+	1850
	Wave Inshore	o	o	o	0
	Wave Offshore	o	+	o	0
Wind	Onshore	+	+	+	3500

Key: + Yes   o No

Table 2c   Norweb's potential for renewable energy

Good news . . . but how good?

The good news is that there is plenty of energy all around us. What is more, no matter how much of it we use, it keeps on replacing itself. This renewable energy is found in natural energy resources in the environment. These are the wind, waves, tides, sun, hydro and biofuel.

Sounds too good to be true? Unfortunately it is. Renewable energy has four main problems:-

1 Unlike the energy in a lump of coal or tank of oil which is in a concentrated form, renewable energy is so thinly spread that it cannot be harnessed very easily or economically.

2 Much is available at the wrong time or not at all. For example, sunlight is not available at night. It is reduced in winter when heat is needed most. There is little wind on a calm day.

3 We cannot replace our present energy system overnight. It could take 20 years to develop renewable energy resources properly.

4 Utilising renewable energy resources may have an impact on the environment.

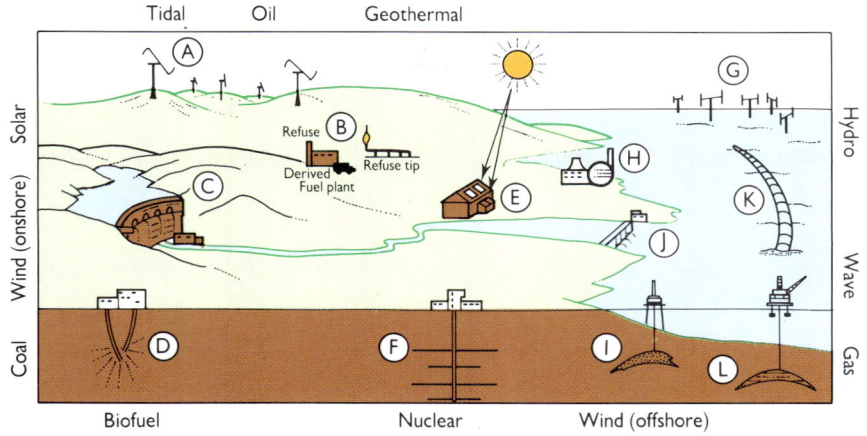

Figure 2.3   Energy resources

Like it or not – it is the law

In 1989 the Electricity Act was passed in the UK. Part of the Act said that all area boards must buy some renewable energy. The first board to study how much energy renewables could supply in its area was Norweb. It studied 12 renewable energy technologies. The study asked three questions:-

1 Was there a resource?

2 Was exploitation technically possible?

3 Was it economic?

Table 2c shows the results of the study. Not all of the resources can be developed for technical, economic and environmental reasons. Even so, Norweb thinks that the renewables can supply 12% of its electricity. Much would come from biofuels and wind power.

Ⓠ Ⓤ Ⓔ Ⓢ Ⓣ Ⓘ Ⓞ Ⓝ Ⓢ
Core

1 Why is energy an important environmental issue?

2 Name three finite or non-renewable fuels.

3 Name three renewable energy resources.

4 Give two problems associated with renewable energy.

5 Which fuels are used in your house for:
(a) heating rooms   (b) cooking   (c) lighting.

6 Copy figure 2.3. Put the energy names next to the correct letters on the diagram.

A
1 What uses most energy in a house?
2 Which was Britain's most important fuel in 1989?

3 How long might Britain's coal last?

4 Which two renewables might help Britain most in 2025AD?

5 Name the four renewables being considered by Norweb.

B
1 What percentage of energy in a house is used by space and water heating?

2 Which fuel will be most important for Britain in the future?

3 What percentage of Britain's energy need in 2025AD does the Department of Energy think could be supplied by renewables?

4 Why is geothermal energy not attractive to Norweb?

Figure 3.1   Tonypandy

Finite energy

unit 3

Some you win ... some you lose

Figure 3.2   The Wistow mine

Look at figure 3.1. Old coal mines like this were not environmentally friendly. They were local eyesores and produced noise, dust, smoke and water pollution. Removing the coal often led to subsidence which damaged houses, roads and fields.

New coal mines should be developed in a way which causes as little harm as possible to the environment.

Some you win

Selby is Britain's biggest single energy find to date. It has five coal seams with reserves of 2000 million tonnes. Only 300 million tonnes will be mined to protect the surface from subsidence.

Lucky?
At Selby the planners have been helped by some good fortune. Having a thick, easily-worked seam of clean coal near the surface at one edge of the coalfield has allowed them to plan a new type of mine. Coal is mined at five deep mines and brought to the surface at a single drift mine.

This has brought several advantages. They are:

ADVANTAGE	REASON
1 Small deep mine sites	No land is needed for rail sidings, coal handling plant or coal stockpiles.
2 Coal brought to the surface more efficiently.	Because continuous conveyors can be used – much better than the stop-go method of winding coal up a shaft.
3 No waste tip or coal washery.	Because the coal is 'clean'.
4 Power stations are the only market.	Three of Europe's biggest – Eggborough, Ferrybridge and Drax are only a few miles away.
5 No greenfield land is needed for the rail yard at the drift mine.	A disused rail yard beside the main line to the power station is available at the drift mine.
6 Screening banks can be built around the mine site.	Using material excavated from the shafts.

British Coal were given planning permission to develop Selby after a public enquiry in 1975 said the coalfield was important for the country.

Some you lose

In the 1970s British Coal found a large new coalfield with 510 million tonnes of coal. Most of it was in Leicestershire. The North East Leicestershire Project (NELP) submitted plans to mine the coal. The Selby system for extraction could not be used because the geology of the coal seams was too complicated. The plan was to mine coal from the three deep mine sites shown in figure 3.3.

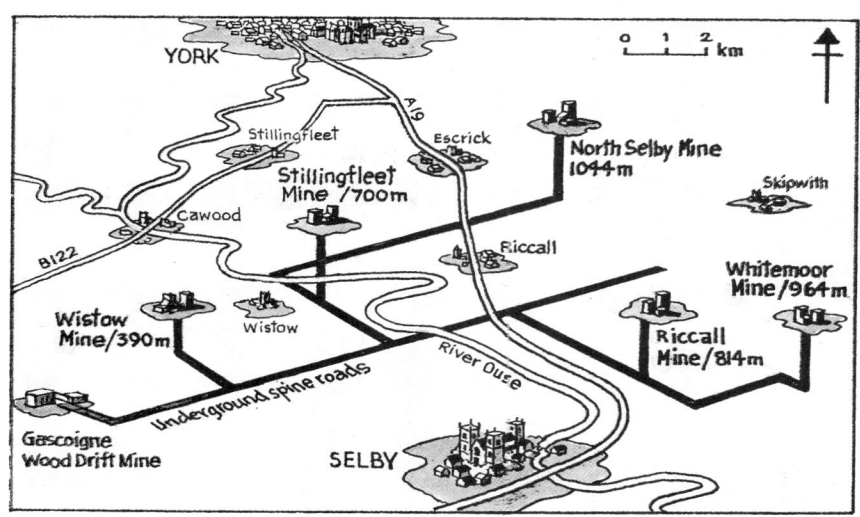

Figure 3.4   The Selby coalfield

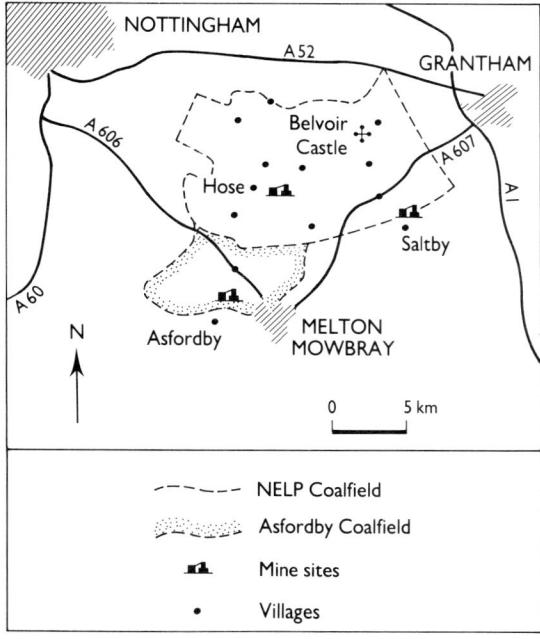

Figure 3.3   The NELP coalfield

Environmental issues

The NELP coalfield lies beneath a rural landscape important for farming, scenic value and wildlife. It was feared that development might alter the character of the landscape. Six environmental issues were considered to be important. They were:

1 Visual	Would the mine sites spoil the local environment?
2 Noise	What noise levels would be produced during construction and operation?
3 Landscape	Would the waste tips and mine towers spoil the rural landscape?
4 Agriculture	Would subsidence affect the drainage network in the fields or damage buildings?
5 Roads	Would they become overloaded with traffic?
6 Traffic	Would extra traffic increase noise and hazards in local settlements?

Public Enquiry

A six month public enquiry in 1984 rejected British Coal's proposal. A main reason for refusal was the creation of large waste tips.

New application

In 1984 British Coal submitted a new application to develop part of the coalfield at Asfordby. Land needed for the mine site and coal tip was reduced from 176 to 141 hectares. This second proposal was passed. The mine is due to be fully operational in 1993.

Core

1 Give three environmental impacts of an old coal mine.

2 Why was the Selby coalfield given planning permission?

3 Why was the NELP coalfield refused planning permission?

4 What percentage of the NELP coalfield was later given planning permission
(a) 25%   (b) 50%   (c) 75%?

5 List three of the main advantages of the Selby system.

A
1 How deep is the Wistow mine?
2 Which is the deepest mine in the Selby coalfield?
3 How long is the underground spine road at Selby?

B
1 How far will coal travel underground from the deepest mine at Selby to the drift mine?
2 Use the sketch, figure 3.4 to draw a map of the Selby coalfield. Use suitable symbols for towns, villages, mines, roads etc.

Big holes

Figure 4.1   Public protest

What message is the poster in figure 4.1 trying to get across?

Opencast coal mining uses very large excavating machinery to dig coal from near the surface of the ground. Unlike deep mining no shaft is sunk, no tunnels dug and only a few people are employed. Most opencast sites operate for between 2–8 years. They are worked day and night. Approximately 15% of Britain's coal is produced from opencast sites. Opencast coal is cheaper to produce than deep mined coal.

Opencasting

There are five stages in operating an opencast coal site. They are:-

Stage	What is removed or replaced	How	By	Then
1	Topsoil and subsoil	Stripped	Scraper	Stored in mounds around site to act as visual screen and noise barrier
2	Overburden	Blasted and dug	Dragline	Placed in previous '**cut**'
3	Coal	Dug	Shovel	Taken from site by lorry
4	Topsoil and subsoil replaced	Carried from mounds	Scraper	Graded and levelled
5	Restoration. Before the land is given back to the owner it is managed for five years by the Ministry of Agriculture. A water supply is laid on, field drains laid, ditches dug, hedges planted or fences built, fertilizer applied and grass seed sown. Finally, the land can be restored to agriculture, recreational use or as a wildlife habitat.			

Figure 4.2   An opencast site

Why the fuss?

Some people living near opencast sites complain that they spoil the quality of the local environment by causing:-

▶ Noise

▶ Dust

▶ Vibration from blasting

▶ Visual pollution

▶ Traffic problems

Some environmentalists also claim that:-

▶ Stored topsoil loses its fertility.

▶ The natural water drainage pattern is destroyed.

When is enough enough?

Figure 4.3 shows the opencast sites that have operated in the Tow Law area from 1953 to 1989. Site 1 started in 1953. Site 42 was at the restoration stage in 1989.

An application to open a new site near the village of Sunniside was made in November 1989. Many local people say that they are fed up with opencasting in the area. They want the application to be refused.

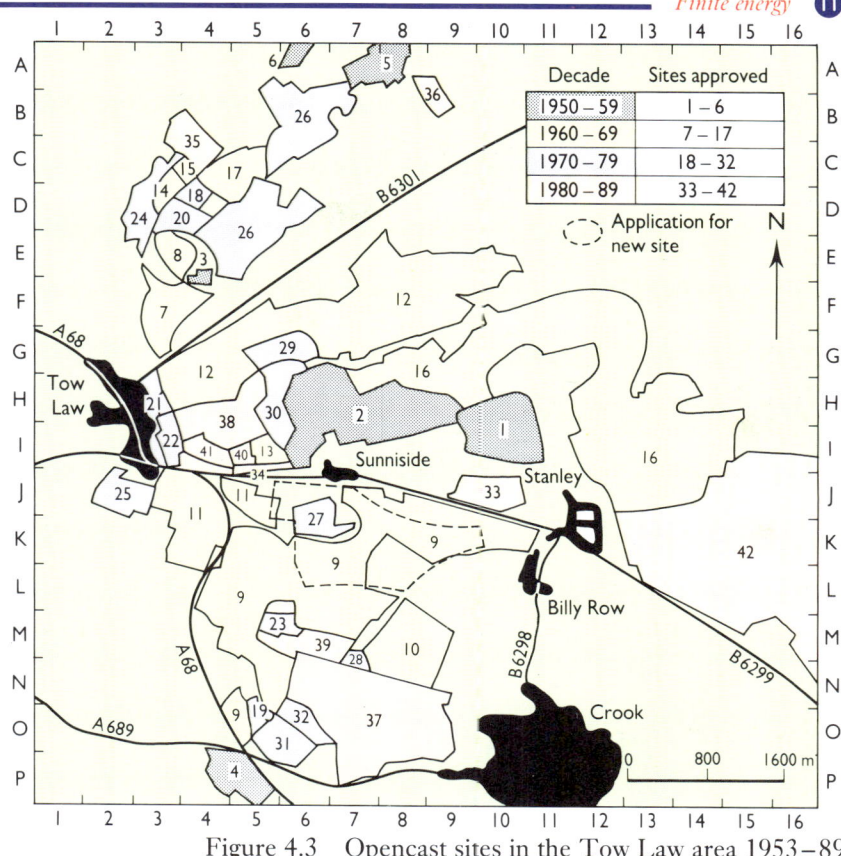

Figure 4.3   Opencast sites in the Tow Law area 1953–89

Core

1 What is mined at an opencast site?

2 List the five stages in the opencast operation.

3 Why is topsoil stored in a mound around an opencast site?

4 List the environmental problems caused by opencasting.

5 What has the land been restored to in figure 4.4?

Study figure 4.3.

6 In which decade was Site 26 worked?

7 Which were the three largest sites worked from 1960–1969?

Class discussion

Should the latest planning application be approved or rejected?

Figure 4.4   A restored site

All at sea
Accidents

Figure 5.1 shows the five stages in bringing oil from under the ground to the consumer. All of these stages affect the environment in some way. Can you think how?

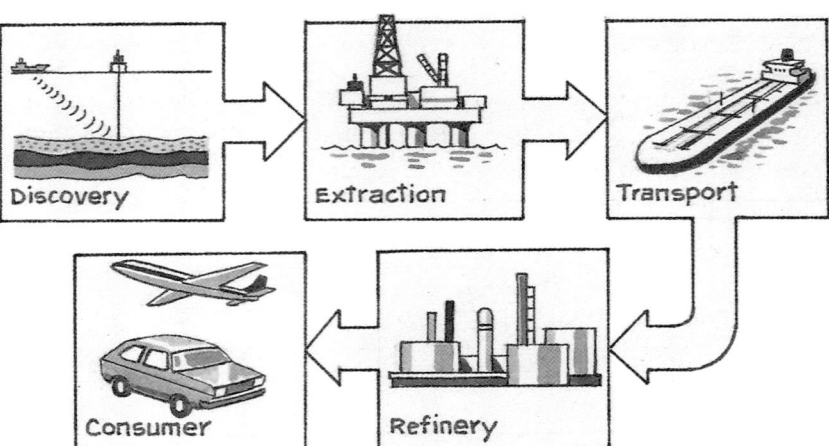

Figure 5.1 Oil stages

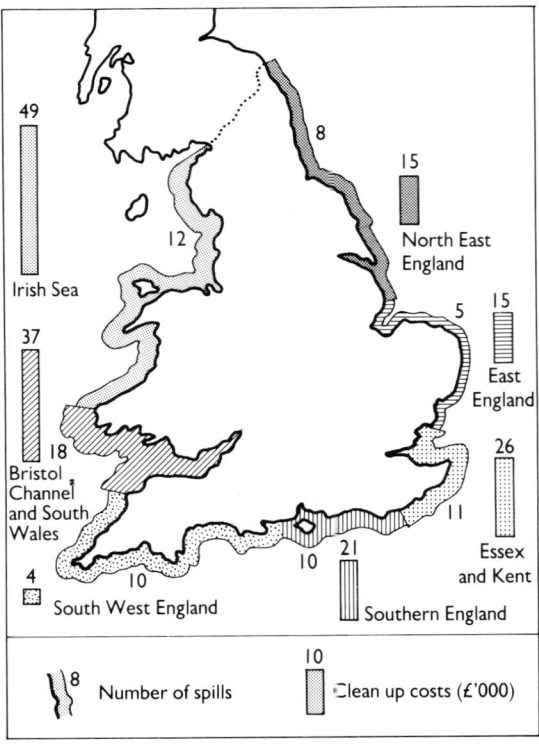

Figure 5.2 Oil spills and clean up costs 1988

Oil technology

Before 1960 oil industry technology could only exploit oil reserves in 'friendly' environments. All the world's oil fields were either:
► On land in warm and hot climates. For example, in the Middle East.
► In very shallow waters in warm climates. For example, off the Louisiana coast in the Gulf of Mexico.

In the 1960s two huge oil finds were made. Both were in very 'harsh' environments. They were:-
► Under the deep, stormy water of the North Sea.
► In the very cold climate of Alaska's ice-bound Arctic Slope.

Old oil technology had no hope of exploiting these massive energy reserves. New techniques were developed. They included the construction of:
► deep sea drilling and production platforms
► underwater and cold terrain pipelines
► large oil tankers

These new technologies have produced oil from harsh environments. Unfortunately, they have also brought environmental problems to them.

Exploiting oil resources in the UK

Extraction stage   In 1988, 166 workers were killed by an explosion on the Piper Alpha production platform in the North Sea.
Transport stage   Pipelines transport about 80% of North Sea oil onshore. There have been no serious environmental problems with pipelines. Most land pipelines are buried underground. They cause local disruption only while being laid. Tankers transport about 20%. Oil spills occur around Britain's coast every year. Table 5a shows the number of spills needing clean-up between 1983–1988.
Refining stage   Thousands of birds were oiled when 150 tonnes of oil spilled from Shell's refinery on the Mersey estuary in 1989. Shell were fined £1 million.

	'83	'84	'85	'86	'87	'88
Oil spills	117	140	139	126	105	120
Costs (£'000)	149	67	107	134	198	217

Table 5a UK spills and clean up costs 1983–1988

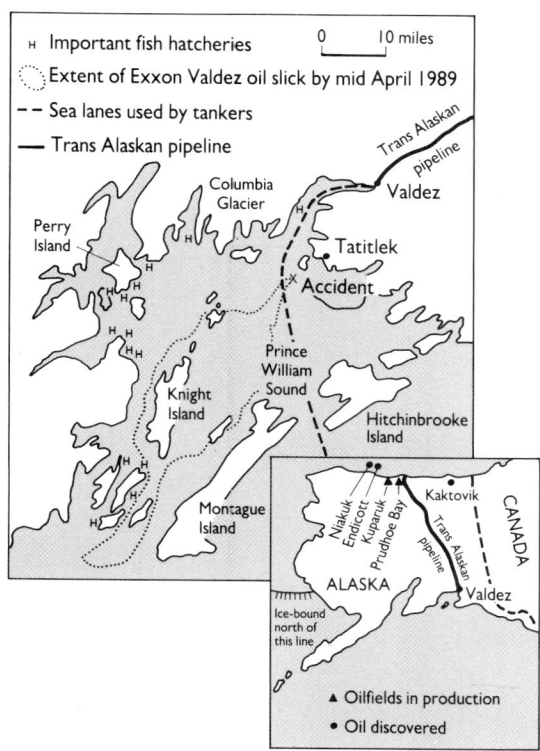

Figure 5.3   America's worst ecological disaster

In the 1970s the route chosen for the TAP caused much controversy. The pipeline was re-routed to avoid conflict with migrating caribou and insulated to stop the hot pipeline melting the permafrost and damaging local ecology.

Better management

All oil spills are a threat to the environment. This is because they can put a lot of oil into a small area in a very short time. The industry needs to be managed in ways which reduce the risk of accidents. For example:

▶ using better navigation equipment on tankers
▶ choosing tanker and pipeline routes carefully to avoid environmentally sensitive areas
▶ better training for people
▶ better tanker design
▶ automatic safety controls at refineries

Exploiting oil resources in Alaska

By 1990 two large oilfields were in production on Alaska's Arctic North Slope. Other oilfields have been discovered. The Arctic coast is ice-bound for most of the year so oil is transported 800 miles to the ice-free port of Valdez by the Trans Alaska Pipeline (TAP).

Oil has brought Alaska jobs and wealth. It provides 80% of the state's income. It has also caused several environmental problems.

Production stage   The 180 000 strong Porcupine caribou herd is one of the last great animal herds to roam North America. The herd breeds in the Arctic National Wildlife Refuge around Kaktovik. Oil companies have found oil at Kaktovik and want to exploit it. They argue that more oil will keep the TAP full and help Alaska's economy. Environmentalists argue that a 1987 agreement between USA and Canada protects the caribou. They fear oil development will reduce the herd by 40%.

Transport stage   Transporting oil from Valdez caused what some people have called America's worst ecological disaster.

Where — Prince William Sound
When — 24.3.1989
How   — Tanker *Exxon Valdez* hit a reef and spilled 10 million gallons of oil.

Ecological damage

Birds   – about 10 million were oiled and died.
Whales   – some probably died.
Salmon   – many migrating salmon ingested oil and died.
Sea otters – oil clogged their fur reducing their insulation against the cold. Many froze and drowned.
Caribou   – some ate oil-covered seaweed and were poisoned.
Eagles   – some ate oiled fish and otters and died.

Core

1 Name two oilfields situated in harsh environments.

2 What technology is needed to bring oil from:
   (a) the North Sea
   (b) Alaska's North Slope?

3 Use table 5a and figure 5.2 and information in the text to briefly describe problems brought by the oil industry to Britain.

4 Why was the Trans Alaska Pipeline re-routed?

5 Give one reason for and one against developing the new oilfield at Kaktovik.

6 Study figure 5.3;
   (a) How far was the tanker off-course?
   (b) Did the slick drift in a SE, SW or NE direction?
   (c) What was the maximum length and width of the slick?

7 Describe three effects of the Exxon oil spill on the local ecology.

8 Suggest three ways to reduce the risk of oil spills.

Slick

An oil slick hits this coastline. How will it be cleaned up? Will it harm any wildlife? Will it reach the tourist beach? What will be the final cost of the clean-up?

Work in pairs. Deal with one oil slick first. Then go on to deal with your partner's oil slick.

Your teacher should guide you through Steps 1, 2, 3 and up to day 4 in Step 4. After that you are on your own.

Step 1　Before you start

1　Fix a photocopy of figure 6.1 and table 6a into your book.
2　Carefully study tables 6b and 6c.
3　You need a dice.

Step 2　Important points to remember

1　You can use <u>more than one</u> clean-up <u>option in many squares.</u>
2　You should be using the absorbent mats in <u>all</u> squares in rows 1–4.
3　You should be using the skimmer in <u>all</u> squares in rows 1–5.
4　You should be using chemicals in <u>all</u> squares where there are no fish, shellfish, birds, seals or otters.
5　You should be using manual labour <u>in</u> all squares in row 1.

Step 3　Some examples

1　In E6 you can only use chemicals.
2　In E5 you can use chemicals + skimmer.
3　In F5 you can only use the skimmer. Why?
4　In E3 you can use chemical + skimmer + mats.
5　In H3 you can use skimmer + mats but not chemicals. Why?
6　What can you use in G1?

Day	Square	Oil	Clean-up option	Oil cleared	Oil left	Cost	Environmental damage
1	B9	500	/	0	500	0	/
2	C9	500					
3	C8						
4	D8						
5							
6							

Table 6a　Environmental assessment

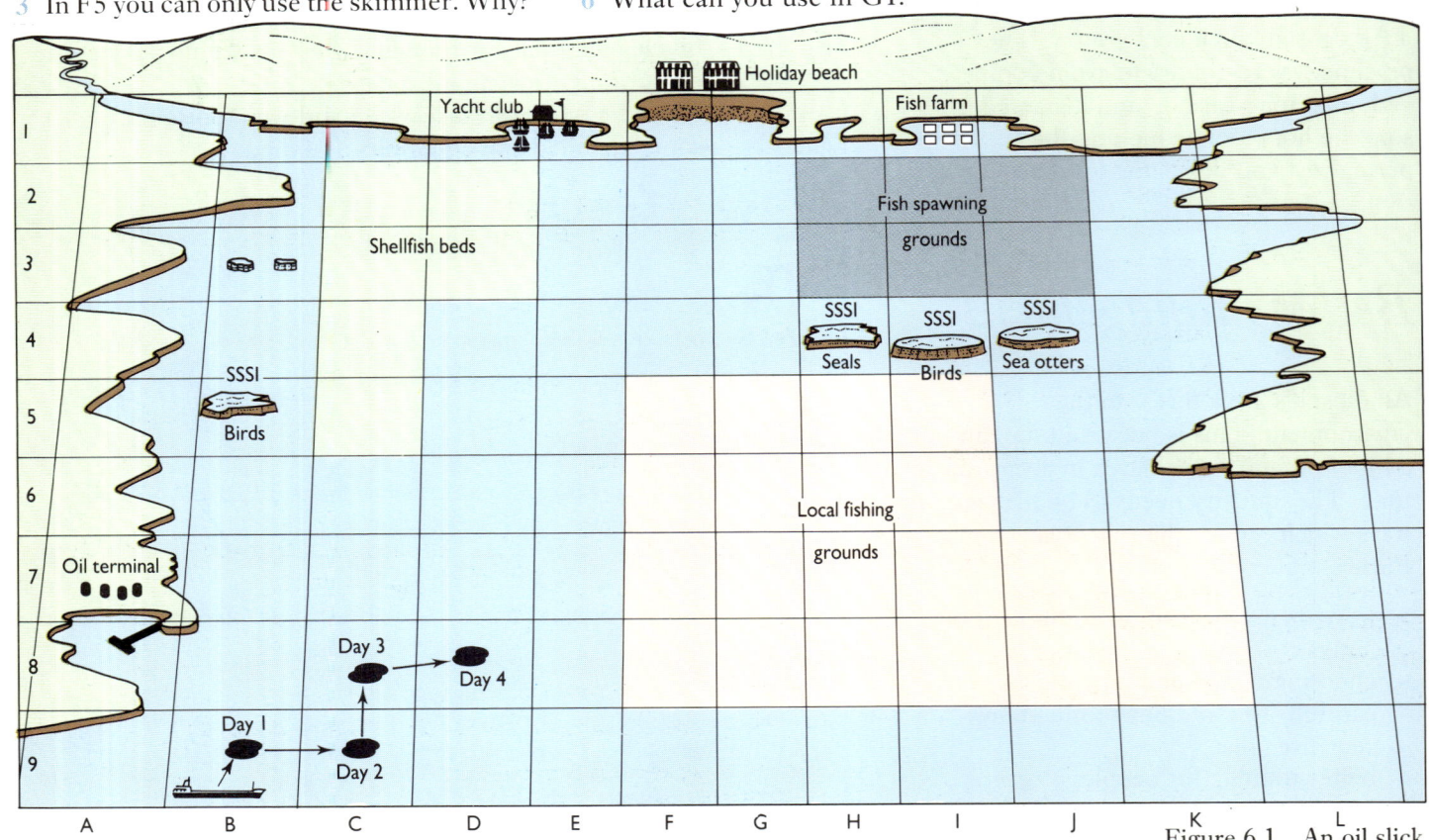

Figure 6.1　An oil slick

Step 4 Situation

Day 1

A tanker approaches the oil terminal. It runs aground on a reef and spills 500 units of oil in square B9. Nothing can be done today. A day is needed to assemble the clean-up options.

Day 2

The slick drifts into square C9. Choose your clean-up options from table 6b. Now, on your copy of table 6a record the:

- option you choose   – in column 4
- amount of oil cleaned up – in column 5
- amount of oil left   – in column 6
- the cost of the day's clean-up – in column 7
- any environmental damage   – in column 8

Look at table 6c for environmental damage.

If the oil slick lands in their square, it
kills
50% of the bird population in that square.
5% of the seal population in that square.
60% of the otter population in that square.
destroys
1 unit of shellfish beds.
1 unit of fishing grounds.
1 unit of fish spawning grounds.
the fish farm.
damages
Yachts – add £20 000 compensation to costs.
Beach – add £50 000 compensation to costs.

Table 6c   Environmental damage

Day 3

The slick drifts into C8. Write the amount of oil still to be cleaned up in column 3. This is the same amount as in day 2 column 6. Now choose your clean-up options again. Now fill in columns 4–8.

Day 4

The slick drifts into D8. Fill in columns 3–8 as before.

Option	Code	Daily oil cleaned up (units)	Cost/day (units)	Use
Skimmer	S	50	10	Needs calm water. Can only be used in rows 1–5.
Chemical dispersants	C	20	10	Toxic. Cannot be used in squares with fish, birds, shellfish, seals or otters. Can be used in all other squares in rows 1–9.
Absorbent mats	A	10	5	Can only be used in rows 1–4 where there is very sheltered water.
Manual labour	M	10	10	Only on the beach. Row 1.

Table 6b   Clean-up options

Day 5

Roll a dice to see the new position of the slick. The rules are:

DICE ROLL	SLICK MOVES
1	one square forward to the left
2	one square forward
3	one square forward to the right
4	one square to the right

Example 1   If you roll a 1 the slick moves into C7
Example 2   If you roll a 4 the slick moves into E8

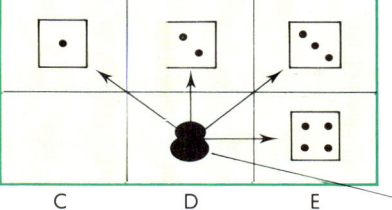

Where would the slick be if you had rolled a 3? a 2?

For a 5 or 6, roll the dice again.
Now plot the new position of the slick on your map. Then fill in columns 2–8

Day 6, 7, 8 etc. Continue to,

- Roll the dice.
- Plot the new position of the slick on your map.
- Choose your clean-up options.
- Make your recordings.

Last day

Keep rolling the dice till the slick enters a square in row 1. The oil is now on the beach. You must still clean all of it up.

Report

Write a short report to describe this oil spill. Include in the report,

- in which square the slick reached the beach.
- the total cost of the clean-up.
- what environmental damage was done.

Making electricity

The tomato-growing giant

Look at figure 7.1. Would you like to open your curtains to this view every day? If you lived at nearby Carlton would you want to sell your house? Would anyone want to buy it? If you had to live in this area which village would you choose to live in? Are jobs more important than views? To answer these questions read on.

Inputs

Drax is a giant with the biggest appetite in the country for coal. This is the fuel the power station burns to generate electricity. Twenty trains feed Drax each day with 1000 tonnes of coal each. Bulldozers stack the coal into stockpiles before it is fed to Drax's furnace to be burned.

Figure 7.1   Drax power station

A room with a view

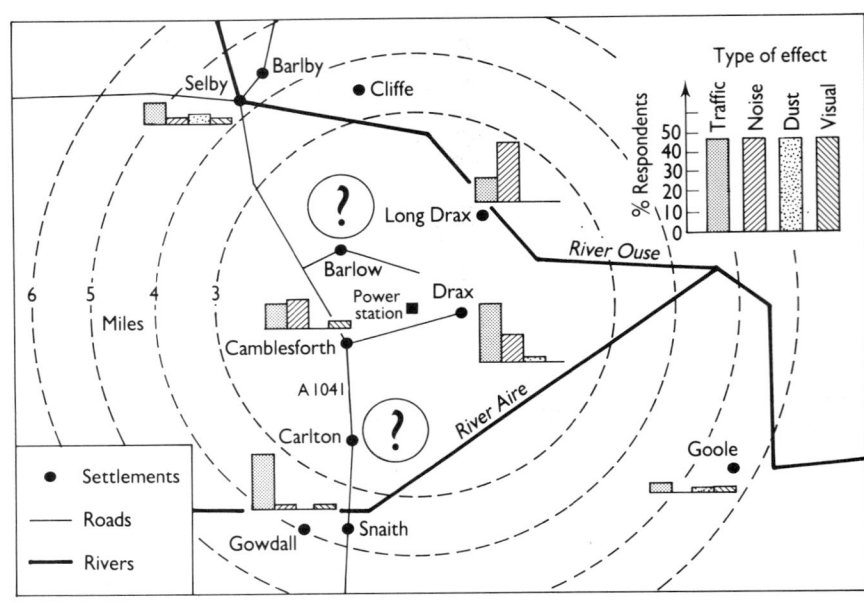

Figure 7.2   Village views of Drax

Outputs

The power station has several outputs. The output people want is electricity. Drax is now Britain's biggest power station. It can produce 4000 MW of electricity. Enough to meet the needs of a city of 4 million people.

Drax also produces other outputs which people do not want. These are the environmental costs of making electricity by burning coal. The environmental costs felt locally are:-

▶ Smoke and dust pollution from Drax's stack, coal store and the trains.
▶ Traffic problems caused by lorries taking ash away for disposal.
▶ Visual impact of a 500 m tall stack and 12 massive cooling towers with vapour 'plumes'.
▶ Noise made by trains and lorries bringing coal and taking ash away.

Benefit	%
Employment	62
Community	16
Trade	10
Wages	7
Housing	5
Services	5
Others	6

Table 7a   Benefits brought by Drax (% of respondents)

Cost	%
Environment	36
Traffic	18
Landscape	12
Community	6
Housing	4
Employers	3
Others	11

Table 7b   Environmental costs of Drax (% of respondents)

Tomatoes?

A power station growing tomatoes? An unexpected benefit from Drax is tomatoes. In the past, all the hot water produced from cooling the turbines used to be wasted up the cooling towers. Now a small amount is recycled by piping it to heat glasshouses which grow 2200 tonnes of tomatoes each year.

Village	Traffic	Noise	Dust	Visual
Barlow	3	24	8	8
Carlton	47	12	11	4

Table 7c   Local feelings in Barlow and Carlton (% of respondents)

Local feelings

A survey was carried out in 1982 to find out how people living in eight local villages thought Drax had affected their environment. In 1982 Drax was a 2000 MW power station with six cooling towers. It now has twelve cooling towers and burns twice as much coal. The results of the survey showed that local people thought there were both benefits and environmental costs.

Table 7a shows the benefits. Table 7b shows the environmental costs. Figure 7.2 shows how local people in six of the villages felt about the four main environmental costs. Table 7c shows the local feeling for the other two villages, Barlow and Carlton.

Core

1 What is Drax? What fuel does it burn? What does it produce?
2 How is coal brought to Drax and ash taken away?
3 Name the four main environmental costs produced by the power station.
4 What is seen as the most important benefit?

A
1 Name the parts of the power station labelled A and B.
2 What is seen as the main environmental cost at:-
   (a) Carlton   (b) Long Drax?
3 Has Drax brought more benefits than environmental costs?
4 So, which village would you choose to live in f you moved to the area? Why?
5 Draw a cartoon sketch like the one in figure 7.3 to show Drax's inputs and outputs.

Figure 7.3   Inputs – outputs

B
1 How is some of Drax's waste hot water recycled?
2 Can you work out if the ash lorries travel;
   (a) North to Selby   (b) South to Snaith?
3 Trace the map, figure 7.2 into your book. Complete it by adding bar graphs for Barlow and Carlton.
4 Can you prove that local people thought Drax had brought more benefits than environmental costs?

Neighbours

Scram! I've enough problems with acid rain!

Figure 8.1   Good neighbours?

Source	SO$_2$	NO$_x$
Power stations	45	15
Industry	17	7
Commerce and houses	6	1
Transport	1	8

Table 8a   Main British sources of acid rain (%)

Figure 8.2   Damage to stonework on Winchester Cathedral

Good Neighbours?

Does the tree in figure 8.1 regard the dog as a good neighbour? Not likely. Many Swedish people feel the same way about some of their European neighbours. They blame them for damaging Sweden's environment. They say things like, 'We do not know what pure rain is anymore' and 'You're waging chemical war on us'.

The issue is acid rain. It is an important regional issue because pollution produced by one country can travel many miles to cause environmental damage in other countries. Such pollution is called **transboundary** pollution.

Causes of acid rain

Normal rain becomes acid rain when it has been polluted by a 'cocktail' of air pollutants. Sulphur dioxide and Nitrogen oxides are the most important ones. Table 8a shows the sources of these two gases in Britain.

The gases are emitted from chimneys and exhausts. Winds can blow them many miles from where they were produced. In the atmosphere they react with sunlight and water vapour and are changed into sulphuric and nitric acid. They then fall to earth as acid rain.

Environmental Damage

Acid rain can cause damage to three main parts of the environment. They are:-

1 Water — Cause — Acid rain washes toxic chemicals from the soil into rivers and lakes.

Effect — Some fish eggs to not hatch. Some young fish are born with deformed backbones. The gills of some adult fish become clogged with aluminium and they suffocate. Sweden has 18 000 acidified lakes. Four thousand of them have no fish at all.

2 Buildings — Cause — Acid rain corrodes stonework

Effect — Many buildings have been damaged throughout Europe. For example Liverpool, Cologne and St. Paul's cathedrals.

3 Forests — Cause — Acid mists attack leaves. Roots take up toxic chemicals from the soil.

Effect — Trees are damaged. Some areas of European forests have been badly damaged. For example in Sweden, Germany and Czechoslovakia.

Culprits and Victims

All neighbouring countries pollute each other. However, some countries do much more polluting than others. There are culprits and victims. Figure 8.4 shows Sweden as both a culprit and a victim.

CHOICES AND SOLUTIONS

Can anything be done?
It is **worth** doing anything?
Do we **want** to do anything?
Table 8b gives some answers to these questions.

Can anything be done?	Is it **worth** doing anything	Do we **want** to do anything?
YES	YES	YES AND NO
✳ Conserve energy	✳ It would cost the UK £1500m to clean up most of its acid rain	✳ September 1986 – government announce programme to spend £600m on cleaning up power station emissions
✳ Burn coal with low sulphur content	✳ This would add £1–3 to your yearly electricity bill	✳ BUT – government won't join '30% Club' to guarantee reducing SO₂
✳ Fit flue–gas cleaning equipment to power stations	✳ But acid rain causes £40 000m damage in the EC every year!	✳ In 1983 Mori Poll – 56% of British public say 'Yes' to extra £1–3 on electricity bill
✳ Develop alternative energy		
✳ Develop nuclear power		

Table 8b   Can we . . . should we . . . do we?

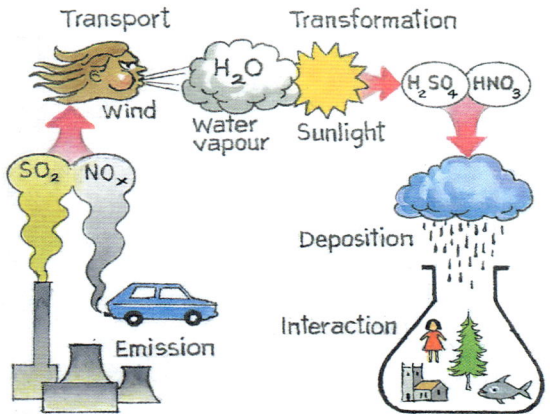

Figure 8.3   Pathway to pollution

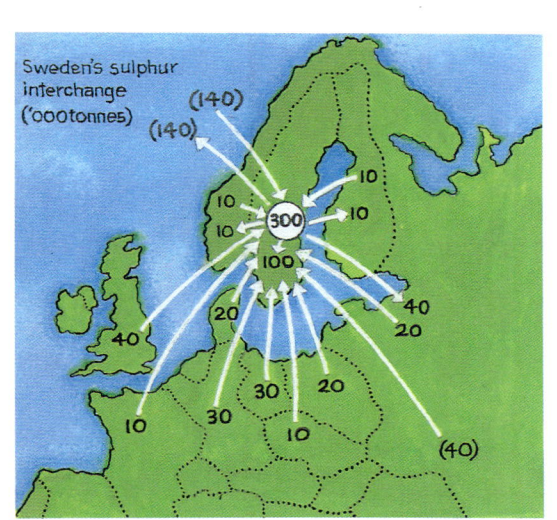

Figure 8.4   Sweden, culprit or victim?

 QUESTIONS

Core

1 Name   (a) The two main gases which cause acid rain.
       (b) The biggest source of these gases.
       (c) The acids which these gases produce.

2 Why is acid rain an important issue?

3 Draw the sketch pathway to pollution (figure 8.3).

4 Give three examples of environmental damage caused by acid rain.

5 Name three things that can be done to reduce acid rain.

A 1 Draw a symbol for a power station, a factory, an office and a car. Write next to each the percentage of (a) SO_2 (b) NO_x they produce.
2 Draw a sketch to show the effects of acid rain on a tree.

B 1 Draw a pie chart to show the contribution to acid rain of power stations, industry, commerce and transport.

2 Using figure 8.4 and an atlas to identify the countries, produce a table to show how much SO_2
(a) Britain and Germany sent to Sweden.
(b) Sweden sent to these countries.

How acid is your rain?

Monitoring acid rain in Britain

The Department of the Environment monitors acid rain in Britain. In 1988 it had a monitoring network of:

▶ nine main sites – they monitor rain every day.

▶ sixty secondary sites – they monitor rain every one or two weeks.

From 1989 only five 'wet' sites have been monitored.

Figure 9.1 shows the results of the monitoring programme for 1987.

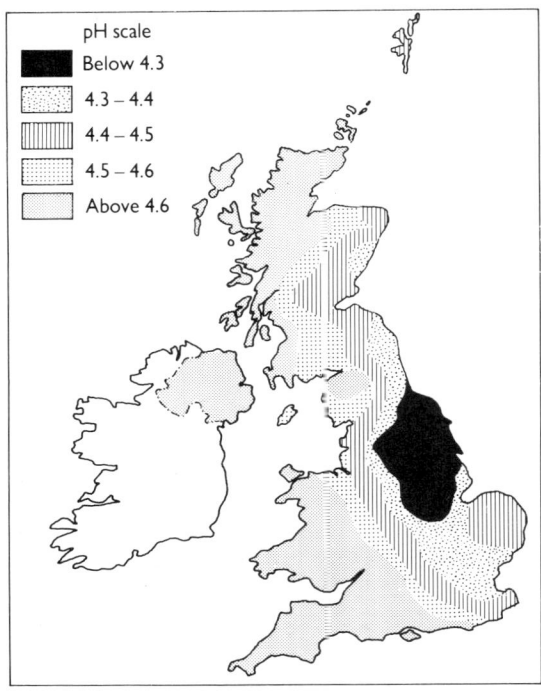

Figure 9.1　Acid rain in 1987

1　According to the map how acid is the rain where you live?

2　How accurate do you think this map can be with only nine main monotoring sites?

3　Why do you think the Yorkshire/Trent Valley/Humberside area has the most acid rain? Want a clue? Look at a map showing the location of Britain's coal fired power stations.

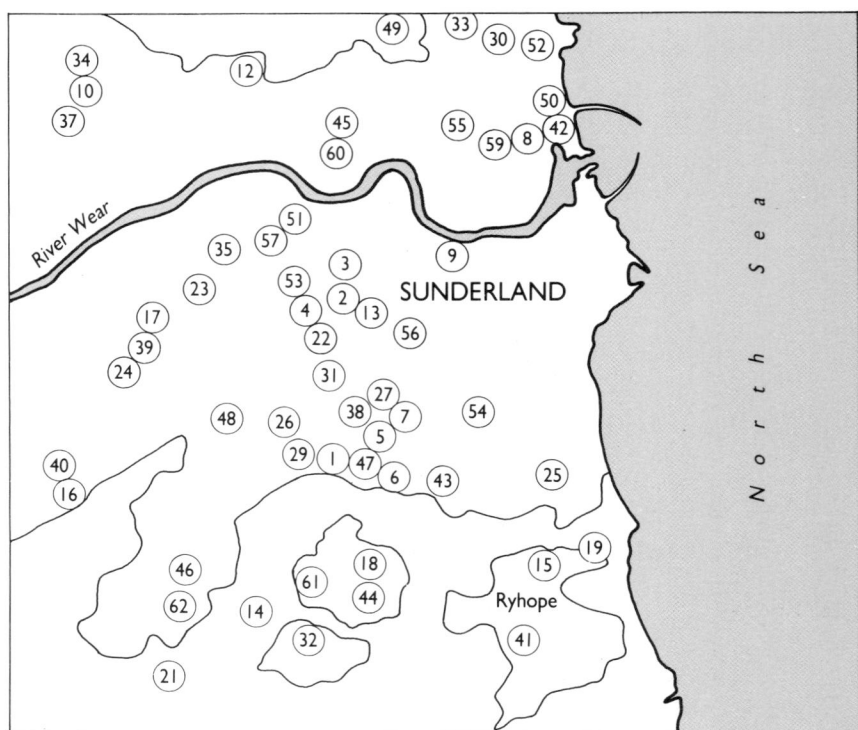

Figure 9.2　Monitoring sites in Sunderland

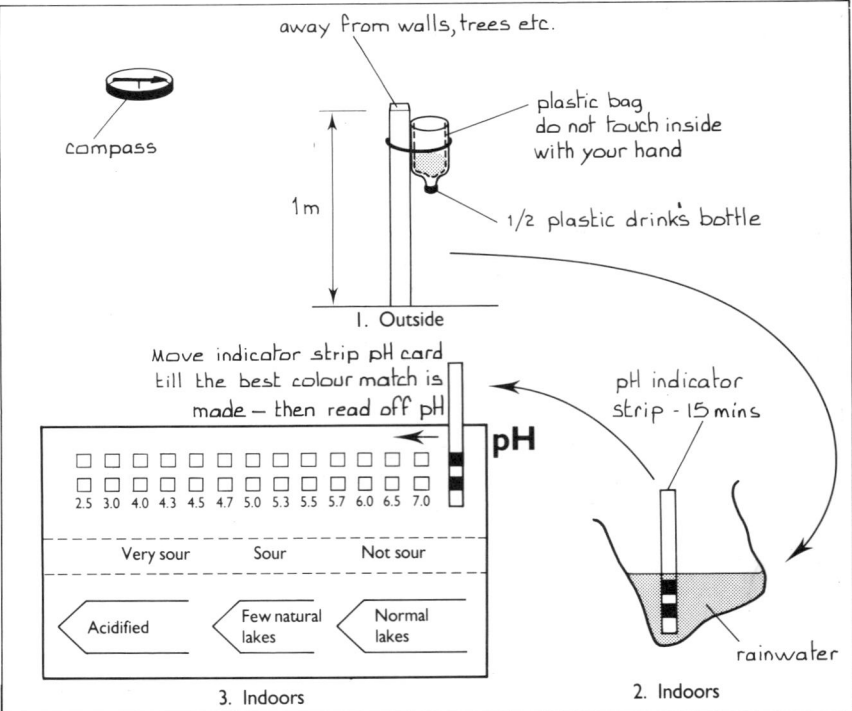

Figure 9.3　How to monitor acid rain

Acid rain in Sunderland

Many pupils have monitored acid rain up and down the country. Pupils in a Sunderland school have monitored the acidity of the town's rain for several winters. Each pupil was responsible for one site. Usually this was their back garden. The sites for winter 1988–89 are shown in figure 9.2. (Some pupils lived outside the town – their sites are not shown on the map.)

This is what the pupils did.

Aim
To investigate acid rain in Sunderland
Equipment
10 pH indicator strips
1 pH card
½ a plastic drinks bottle
10 plastic bags
1 compass
Method
1 Set up the equipment as shown in figure 9.3.
2 Collected a rain sample over 24 hours.
3 Took a compass reading of the wind direction during the rain.
4 Placed the indicator strip in the rain for 15 minutes.
5 Read off the pH value from the pH card.
6 Recorded their results for both pH and wind direction.
7 Took 10 rain samples.
Results
Table 9a shows how the pupils recorded their results.

Treatment of results
The results could have been treated in several ways. The Sunderland pupils:

1 Calculated the average pH for each site.
2 Calculated the average pH for the town.
3 Drew a map to show the acidity of rain at each site in Sunderland.
4 Calculated the average acidity of rain brought by the eight main wind directions as shown in table 9b. They also drew a diagram to represent this data.
5 Made a frequency table of the results as shown in table 9c.
6 Looked for evidence of damage done by acid rain to stonework, streams and plants in the town.

Wind direction	pH values
South East	5.6, 4.9, 5.2, 5.2, 6.0, 5.5, 3.8, 5.6, 5.3, 5.5 5.5, 5.0, 5.3, 3.3, 5.4, 4.0, 5.7, 3.2, 3.9, 5.0 7.0, 3.3, 3.6, 6.0, 3.9, 3.2, 6.7, 5.5, 5.4, 6.5 4.0, 6.0
	Average = 5.1

Table 9b Windborne acid rain

Table 9a Sunderland results

Site No.	Samples																			
	1		2		3		4		5		6		7		8		9		10	
	pH	W	pH	W	pH	W	pH	W	pH	W	pH	W	pH	W	pH	W	pH	W	pH	W
1	5.6	SE	5.5	SW	4.0	SW	4.7	NE	6.5	S	6.5	SW	6.0	N	3.5	NE	4.0	SW	5.5	SE
2	3.2	E	3.8	E	5.7	NE	5.5	S	4.5	S	5.3	W	5.0	NE	4.5	E	5.2	N	4.8	NE
62	5.6	SW	5.6	NW	3.4	SW	5.8	SW	5.6	NE	6.1	NE	6.5	SW	6.0	SE	5.8	SW	5.7	S

pH	Frequency	Total
3.0–3.9	⳽ II	
4.0–4.9	⳽ ⳽ ⳽ I	
5.0–5.9	⳽ ⳽ ⳽ ⳽ ⳽ ⳽ ⳽ III	
⩾ 6.0	II	

Table 9c Acid rain frequency table

So . . . how acid is your rain?

Devise a method to monitor acid rain in your town. It should involve everyone in the class. Carefully plan how you will carry out this task. Will you monitor rain over a year, a winter, a month, or even a single rain storm? How will you record and present your results? You can use the Sunderland ideas or you can develop your own.

So . . . how acid is your rain?

A grey area

Not black, not white

Nuclear power is a very controversial issue. Some people are for it. Some are against it. Most people make their minds up without knowing much about it. It is a grey area.

Chernobyl

The world's worst nuclear accident happened in 1986. An explosion at the Chernobyl nuclear power station near Kiev in the USSR hurled dangerous radiation into the atmosphere. Very high levels of radiation can kill people. Lesser amounts of radiation may cause cancers in later years. Very low levels are most probably harmless.

Local effects

- ▶ 31 station workers received very high levels of radiation and died within months.
- ▶ 60 000 people were evacuated from the town of Pripyat, 5 km away.
- ▶ Within 30 km of Chernobyl 135 000 people were evacuated.
- ▶ No one can be certain how many local people might die in the future as a result of this accident. One Russian estimate is 5000 deaths over the next 50 years.

Wider effects

Winds blew the radiation cloud over most of Europe.

- ▶ *Sweden* Sweden was particularly affected. Rain washed radiation into the soil. Grass roots took up the radiation. Grazing animals ate the contaminated grass.
- ▶ *Switzerland* Swiss people received between 210–1980 **millirems** of normal radiation in 1986. Chernobyl exposed them to another 5–200 millirems.
- ▶ *France* French doctors were not very interested in the accident. They thought it might cause a few extra deaths in France in future years. They said that lung cancer caused by smoking kills 30 000 French people every year. Nuclear power stations produce about 75% of French electricity. Most French people are happy with nuclear power.
- ▶ *Britain* Farmers in Wales and the Lake District were not allowed to sell their sheep for many months after the accident because the animals had eaten grass contaminated with radiation.

The International Committee on Radiological Protection monitored this incident. They estimated the accident might eventually cause;

- ▶ 4600 cancers in Europe
- ▶ 2400 of these could be fatal.

This accident showed how air pollution from one country can travel many miles to affect other countries.

Figure 10.1   The Chernobyl nuclear accident

	%
Natural	
radon	47
thoron	4
internal	12
gamma rays	14
cosmic rays	10
total	87
Man-made	
medical	12
fallout	0.4
occupational	0.2
nuclear discharges	0.1
other	0.4
total	13

Table 10a   Yearly exposure to radiation

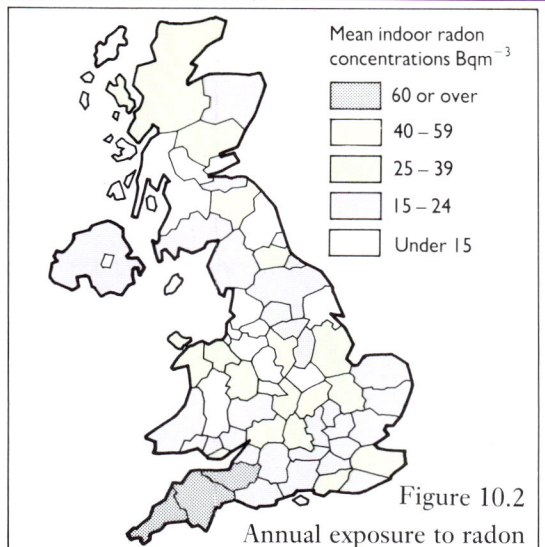

Figure 10.2
Annual exposure to radon

Mean indoor radon concentrations Bqm⁻³
- 60 or over
- 40 – 59
- 25 – 39
- 15 – 24
- Under 15

Nuclear power in Britain

Nuclear power stations produced 20% of Britain's electricity in 1988. The National Radiological Protection Board monitors exposure to radiation in Britain. Table 10a shows the results. **Radon** caused nearly half the total radiation received in Britain.

Radon gas is produced naturally underground. It bubbles up out of the ground. In open air it is diluted. Higher levels can collect in houses. In 1987 the average radon concentration in British houses was 21 **becquerels** per cubic metre. Houses may be at risk if the radon level is more than 400 becquerels per cubic metre. The Board estimated that 20 000 houses had radon levels above this. Figure 10.2 shows most of them were in south west England.

Below are some arguments for and against nuclear power:-

For

► Britain's nuclear industry has a very safe record.

► Radiation from natural sources is far higher than from nuclear power.

► Nuclear power does not produce acid rain or greenhouse gases.

► Fossil fuels will soon run out.

► It produces cheap electricity.

Against

► Nuclear power can be very dangerous.

► An accident like Chernobyl affects other countries.

► The real cost of nuclear electricity is much higher than coal-fired electricity.

► Nobody knows what to do with the dangerous high-level waste.

► Renewable energy technologies are a much safer way to generate electricity.

Core

1 Copy and unscramble
A nuclear power station produces ttleecriciy. It also produces taidraion which can be blown to other countries by the diwn.

2 How far is Chernobyl from London?

3 Copy and correctly label figure 10.3.

4 Write down what you think are the three main arguments for and the three against nuclear power.

5 Use an atlas to name the counties most affected by radon in Britain.

Class survey

Ask the questions opposite and record your results.

Class discussion

(a) Is nuclear power a good way to produce electricity?
(b) Should the EC ask France to close its 53 nuclear reactors?

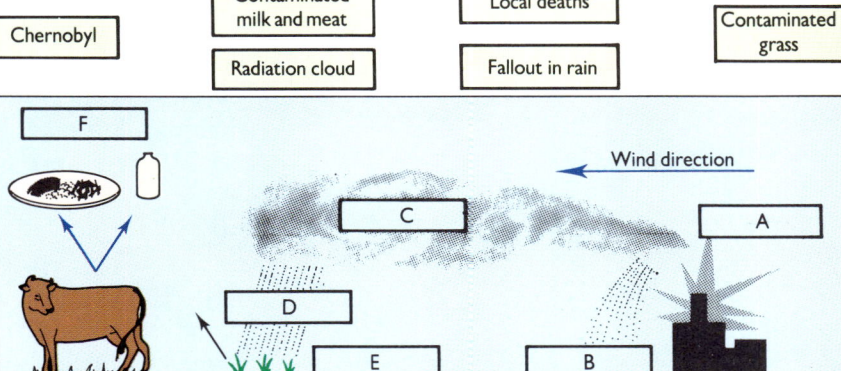

Figure 10.3   Radiation pathway

Policy	Yes	No	Don't know
Build more nuclear power stations			
Keep only existing ones			
Close all nuclear power stations			
A nuclear power station is to be built 10 miles from your house. Would you: Try to stop it Live with it Move house			

Figure 11.1   Denmark's Ebeltoft windfarm

Renewable energy

unit 11

Probably the best in the world

Name of the farm	Equipment
1. Dansk	10 × 55 kW
Oddesund, North	10 × 55 kW
Oddesund, South	24 × 95 kW
2. Tønder	9 (various)
3. Ebeltoft	16 × 55 kW
	1 × 100 kW
4. Ærø	11 × 55 kW
5. Fanø	13 × 55 kW
6. Aale	10 × 75 kW
7. Ranum	14 × 75 kW
8. Sydvestmors	10 × 75 kW
9. Hasle	10 × 99 kW
10. Tændpibe	30 × 75 kW
Velling Mærsk I	34 × 90 kW
	2 × 200 kW
VellingMærsk II	29 × 225 kW
11. Hollandsbjerg	30 × 130 kW
	2 × 300 kW
12. Ryå	20 × 99 kW
	3 × 200 kW
13. Nørrekær Enge I	36 × 130 kW
Nørrekær Enge II	42 × 300 kW
14. Torrild	15 × 150 kW
15. Syltholm I and II	25 × 400 kW
16. Kyndby	21 × 180 kW
17. Masnedø	5 × 750 kW
18. Haslev	6 × 75 kW
19. Klinteby	4 × 250 kW

Table 11a   Danish windfarms 1990

You know the TV advert which claims it is 'probably the best lager in the world'. Carlsberg is the lager but which country brews it?

The same country probably makes the best windmills in the world. These windmills are also called wind turbines or **aerogenerators**. They convert power in the wind into electricity. Denmark probably has more experience with windpower than any country. More than 6000 Danish windmills have been exported to windfarms in California.

Windpower in Denmark

Denmark has no coal and no mountains or rivers for hydro schemes. The country has a limited amount of oil and natural gas. The Danes recognise that the wind is an important source of renewable energy and have begun to develop windpower.

First, they drew up a list of the best technical sites for windfarms. Then they looked at how each site might conflict with other land users. If there were serious conflicts the site was eliminated from the list.

By 1990 Denmark had more than 2500 windmills. Most were in small groups run by the utilities or owned privately or by cooperatives of 50–100 families. There were 19 windfarms. Windpower generated 2% of Denmark's electricity in 1989 and more windfarms are planned. Denmark aims to produce 10% of its electricity from the wind by 1995.

At first local people were worried about having a windfarm near their homes. They were concerned about noise and possible devaluation of their houses. However, windfarms now seem to have been accepted, although large windfarms like those in California are not acceptable in the Danish landscape.

Windpower in Europe

Western Europe has much potential for windpower. The best areas are shown in figure 11.3.

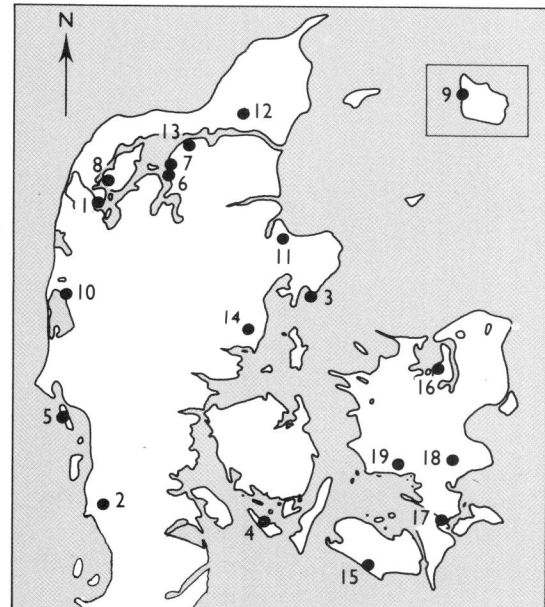

Figure 11.2   Windfarms in Denmark 1990

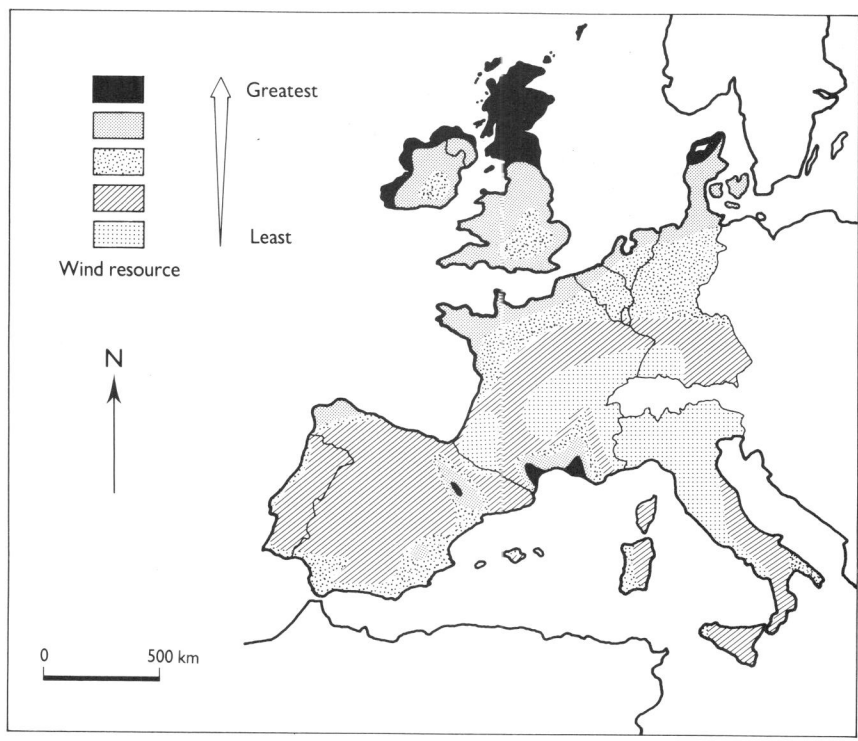

Figure 11.3   The best areas for windfarms in Europe

Windpower in California

California has three wind energy centres. These are mega-windfarms. Table 11b and figure 11.5 show why. They produce 1% of California's electricity.

Figure 11.4   Denmark's Masnedø windfarm

Figure 11.5   A Californian wind energy centre

Centre	No. of turbines
Altamont	6900
Tehachapi	4500
San Gorgonio	3900

Table 11b   Wind energy centres in California

Ⓠ Ⓤ Ⓔ Ⓢ Ⓣ Ⓘ Ⓞ Ⓝ Ⓢ

Core

1 Give two other names for a windmill.

2 What energy does a windfarm use?

3 Is this form of energy finite or renewable?

4 A group of windmills is called a . . . .

A **1** Name windfarm number 3 on figure 11.2.
2 How many windmills make up this windfarm?

3 What do you think is the main difference between a Danish and a Californian windfarm?

B **1** Where are almost all Danish windfarms located?
2 Compare the Danish windfarms in figures 11.1 and 11.4 with the Californian windfarm.

3 Which European countries have the best areas for windpower?

Public trial

Windpower in Britain

Windpower is being considered in Britain. It has generally been accepted in Denmark. Will it be accepted in our country? Several trial wind turbines have been built. For example at Carmarthen Bay, Richborough and in the Orkneys.

The first windfarms

In 1988 plans were announced to build three demonstration windfarms at:
► Cold Northcott in Cornwall
► Capel Cynon in Wales
► Langdon Common in County Durham
(In 1990 the Langdon Common site was abandoned because the foundation conditions were found to be unsuitable.)

Each windfarm will have 25 wind turbines and can produce enough electricity in a year for 5000 people. After measuring windspeeds, the environmental impact of the windfarms will be assessed. Talks with local councils and environmental pressure groups will then be held. Lastly, applications to build the windfarms will be made. The aim of building these sites is to test the public's attitude to windpower in Britain. Some possible environmental impacts of a windfarm are shown in figure 12.2.

Figure 12.1   Photographic impression of the Cold Northcott site

Windpower in Northumberland

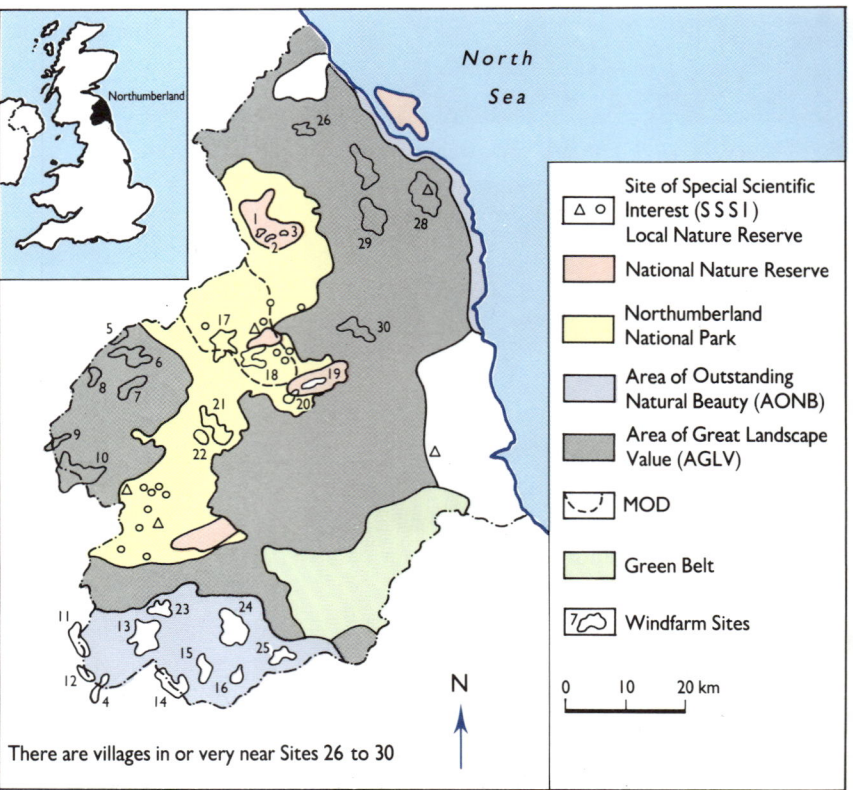

Figure 12.3   Good windfarm sites in Northumberland

Figure 12.2   Possible environmental impacts

A study in 1989 chose the 30 best technical sites for windfarms in Northumberland. The county is very beautiful. Most of it receives some environmental protection for landscape or wildlife. Figure 12.3 shows the sites and the protected areas.

Will windfarms be built at any of these sites? It is too early to say. Some sites probably have no chance. A few might be possible. Northumberland uses about 1200 Gigawatt hours (GWh) of electricity each year.

Core

1 Name the sites for Britain's first two windfarms.

2 Draw the six environmental impacts shown in figure 12.2 and write these comments next to the correct sketch:-
    – Spoil the view or attract visitors?
    – Possible interference.
    – Possible interference.
    – Danger from blade snap.
    – Swishing may be heard up to 300 m away.
    – Possible danger to migrating birds in bad weather or at night.

Site	GWh per year
1	22
2	7
3	7
4	40
sub total	**76**
5	20
6	69
7	49
8	25
9	35
10	104
11	40
12	20
13	119
14	49
15	44
16	35
sub total	**609**
17	26
18	80
19	39
20	3
21	45
22	26
23	19
24	42
25	61
sub total	**341**
26	23
27	47
28	71
29	41
30	18
sub total	**200**
Total =	**1226**

Table 12a   Estimated electricity generation

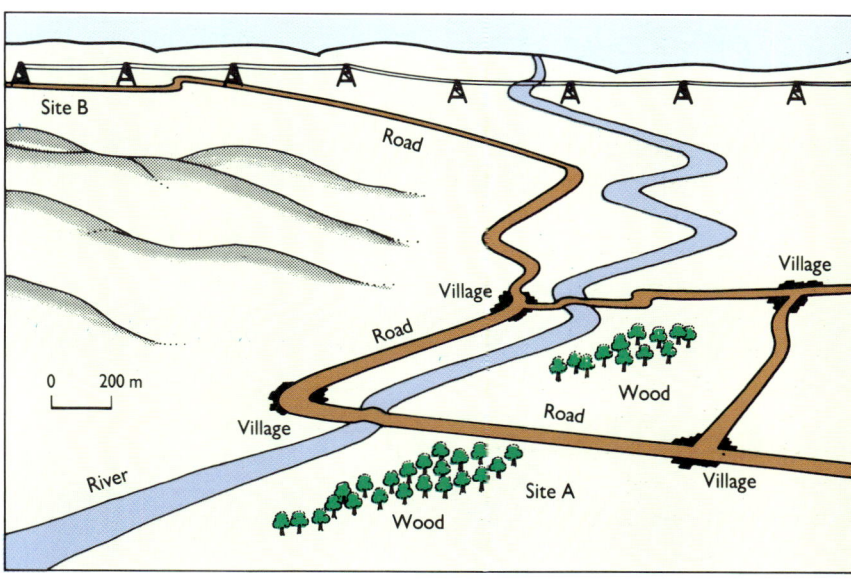

Figure 12.4   Which is best?

3 Study figure 12.4.
  (a) Which site is the best site for a windfarm?
  (b) Give three reasons for your answer.

4 When can a windfarm not produce electricity?

5 Give (a) an advantage   (b) a disadvantage of a windfarm over a thermal power station for producing electricity.

A Study figure 12.3.
  1 What environmental problems face windfarms at sites 1, 5, 13 and 25?

2 When it is built how do you think the Cold Northcott windfarm will affect the environment?

B Study figure 12.3 and table 12a.
  1 Why do sites 1, 2 and 3 probably have no chance of being built?

2 Select three sites for windfarms in Northumberland; choose site 4, one site from sites 5–10 and one site from sites 11–14.
How much of Northumberland's electricity needs can your three windfarms supply?

What a waste?

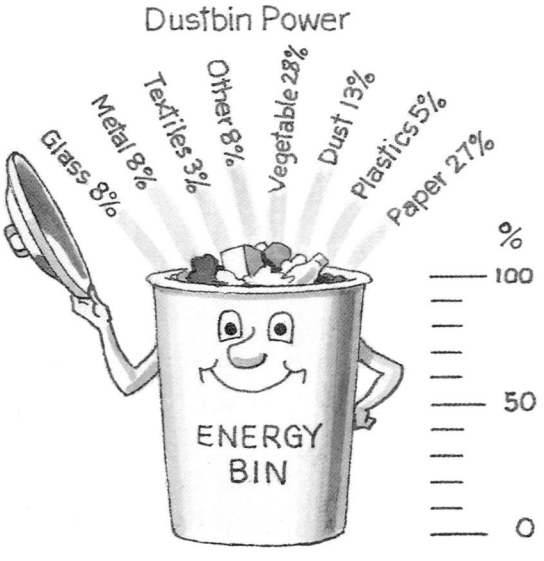

Dustbin Power

Glass 8%
Metal 8%
Textiles 3%
Other 8%
Vegetable 28%
Dust 13%
Plastics 5%
Paper 27%

ENERGY BIN

%
100

50

0

Figure 13.1   Dusty Bin

Dusty Bin

Dusty is looking pretty happy in figure 13.1. He is happy because he knows that what you are throwing away is worth millions of pounds. Just look at what is going in. Most of it will burn. If it will burn it can produce heat. Dusty knows that 2.5 dustbins full of 'rubbish' have the same heat energy as 1 bag of coal. Don't let him hear you call it rubbish.

Dustbin Dan

Dusty is not so happy in figure 13.2 because Dustbin Dan is dumping the savings. Yet this is what is done in Britain. Every year we;

▶ create 30 million tonnes of rubbish
▶ this rubbish has an energy value of £400 million
▶ it could supply heat for 2.5 million families
▶ we spend £255 million dumping rubbish in holes in the ground
▶ 90% of our rubbish goes into holes in the ground

Crazy?

Environmental Impacts

A hole in the ground for dumping rubbish is called a **landfill site**. The impacts landfill sites can have on the environment are;

▶ smell
▶ visual
▶ attracts vermin like rats
▶ attracts scavenging birds like gulls which can spread disease when they fly away.
▶ Danger. Bacteria in the site turns some rubbish into **methane** gas. This gas can build up and might explode. There have been several explosions in the last year or two. In 1988 a report from HM Inspectorate of Pollution said there were as many as 1300 landfill sites with methane building up in them.
▶ Chemicals can be leached from the site by rainwater.

DUSTBIN DAN'S VIEW

WASTE OF A NATION...

Figure 13.2   Dustbin Dan

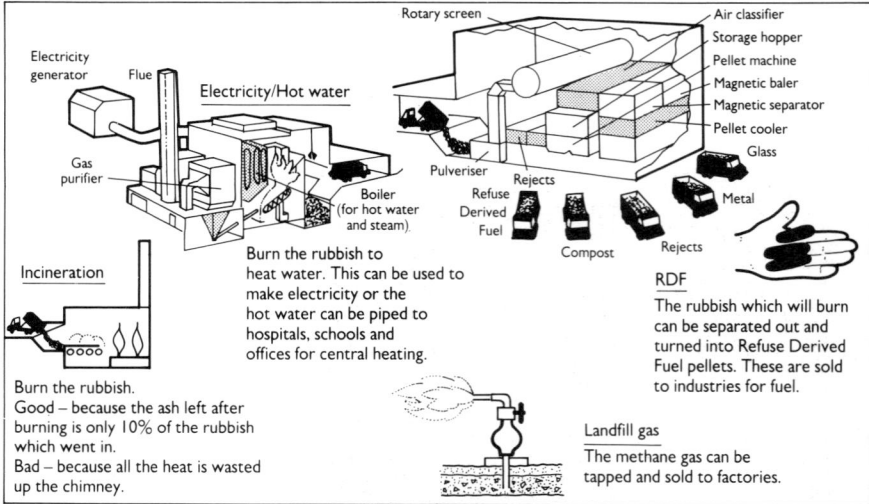

Rotary screen
Air classifier
Storage hopper
Pellet machine
Magnetic baler
Magnetic separator
Pellet cooler
Glass

Electricity generator
Flue
Electricity/Hot water

Gas purifier

Pulveriser
Rejects
Refuse Derived Fuel
Compost
Rejects
Metal

Boiler (for hot water and steam)

Incineration

Burn the rubbish to heat water. This can be used to make electricity or the hot water can be piped to hospitals, schools and offices for central heating.

RDF
The rubbish which will burn can be separated out and turned into Refuse Derived Fuel pellets. These are sold to industries for fuel.

Burn the rubbish.
Good – because the ash left after burning is only 10% of the rubbish which went in.
Bad – because all the heat is wasted up the chimney.

Landfill gas
The methane gas can be tapped and sold to factories.

Figure 13.3   Alternative rubbish disposal

Alternatives

What can be done? Figure 13.3 shows four other ways to get rid of rubbish.

Should we?

In 1988 London's costs for rubbish disposal were;

Method	Cost (£/tonne)
By river	25
By road	15
By rail	20
Edmonton incinerator	10.5

Are we?

There are a number of places in Britain where rubbish is recycled.

Electricity/ Hot water	RDF pellets	Landfill gas
Edmonton	Eastbourne	London
Coventry	Grimsby	London Brick
Sheffield	Newcastle	Company
Jersey	Doncaster	
Nottingham	Liverpool	
	West Bromwich	

Country	Percentage of waste recycled into energy
Belgium	6
Britain	6
Denmark	75
France	21
Germany	27
Holland	18
Italy	4
Luxembourg	76
Sweden	50

Table 13a   The European waste league

Figure 13.4   A landfill site

But?

As table 13a shows, Britain comes nearer the bottom of the European rubbish recycling league than top.

Feeling Guilty?

A recent survey asked people what they thought about recycling rubbish.

▶ 43% said they felt guilty about throwing away so much rubbish.
▶ 89% said their local council should do more to recycle rubbish.
▶ 94% said they would sort their rubbish at home if the council would provide suitable facilities.

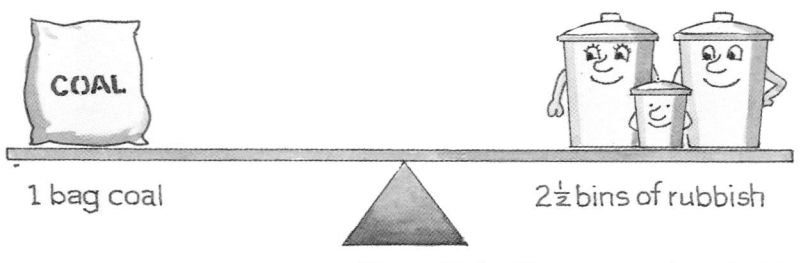

Figure 13.5   The energy value of rubbish

Core

1 What is a landfill site?
2 What percentage of our rubbish is dumped in landfill sites?
3 Why is dumping rubbish a waste?
4 Name three environmental impacts caused by landfill sites.
5 Hold a class survey. Ask these questions and record the results Yes or no.
  (a) Do you feel guilty about throwing away so much rubbish?
  (b) Should your local council do more to recycle rubbish?
  (c) Would you be prepared to sort your rubbish at home?

A 1 Draw a sketch to show how much energy there is in 2.5 dustbins full of rubbish.
2 Draw a sketch to show one method of recycling rubbish.
3 Which is the cheapest way for London to get rid of its rubbish?
4 Name a town which makes,
  (a) electricity   (b) RDF pellets from rubbish.
5 Which country recycles most of its waste into energy?

B 1 Draw a sketch of Dusty Bin.
  Fill him up with the correct percentage for each type of rubbish. Indicate which rubbish will burn.
2 Draw a simple sketch to show why incineration is better than landfill but not as good as energy recovery for district heating.
3 Write a short paragraph to explain the economics of refuse disposal in London.
4 Make a rank order list of the European waste to energy league.

Just how green? *What is it?*

The answer to the engineer's question is the tide. The gravity pull of the moon moves millions of tonnes of water up and down Britain's beaches twice a day to produce high and low tides. That is a lot of moving water producing a lot of energy.

Around most of the country the tidal range, the vertical height between high and low tide, is about 4 metres. However, the funnel shape and sloping bed of the Severn estuary concentrates the incoming tide to give a tidal range of 11 metres between Cardiff and Bristol. The Severn Bore? Not some corny joke teller from Bristol. It is the name of the tidal wave which surfers ride up the estuary. The barrage will be called the Severn Tidal Barrage.

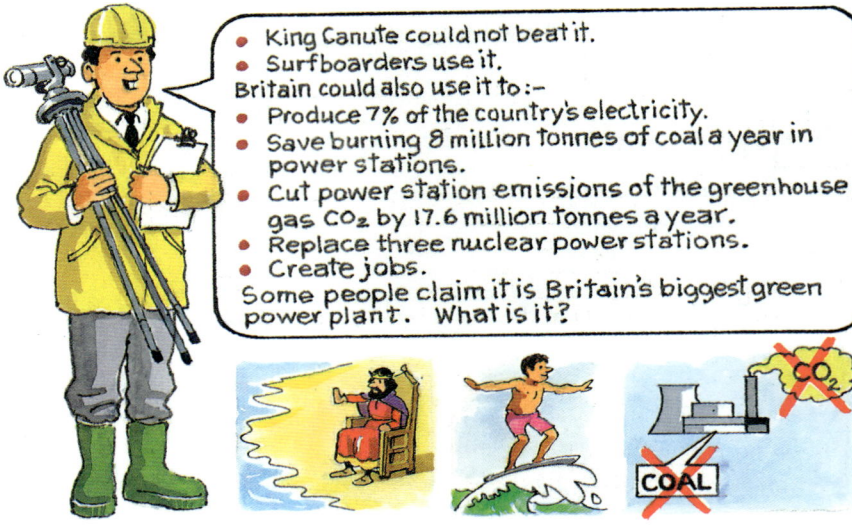

Figure 14.1   What is it?

Britain's biggest green power plant?

The Severn estuary is the second best site in the world for tidal power (the Bay of Fundy in Nova Scotia being the first). Engineers have designed a barrage which could be built across the estuary to harness some of the tide's power. Figure 14.2 shows how the scheme would work.

There are four stages:
1 The incoming flow tide passes up through the sluices.
2 At high tide the sluices are closed.
3 Water on the seaward side of the barrage then ebbs back to low tide. This creates a 'head of water' like an HEP station.
4 Water from behind the barrage now ebbs back to the sea through 216 turbines which generate electricity.

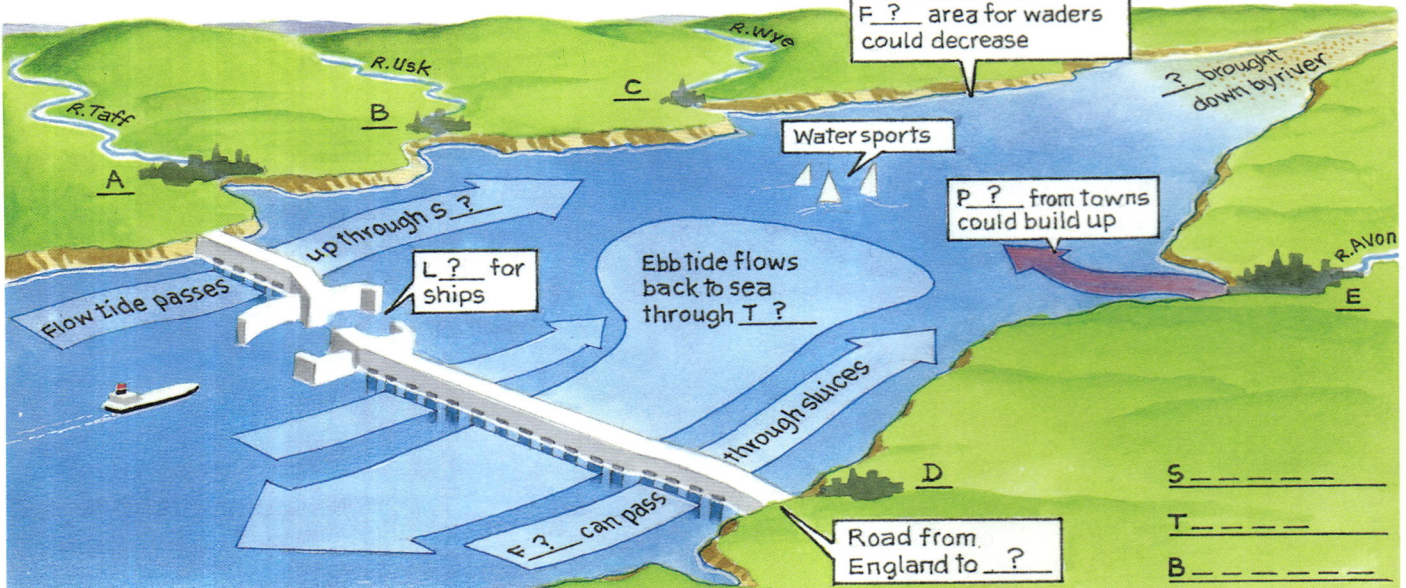

Figure 14.2

WULFRUN COLLEGE
SCHOOL OF SCIENCE

Changes

Building a barrage would cause four physical changes to the estuary;

1 Water level changes.
The new high tides would not be as high, the new low tides not as low as normal.

2 Water flow.
The speed and scouring force of water flowing out of the estuary would be reduced.

3 Sedimentation.
If the scouring force is low, sediment carried downstream by the rivers could build up behind the barrage.

4 Physical.
A barrage would cut off upstream ports from the sea.

Environmental Impacts

▶ *Shipping.* A barrage would cut ships off from upstream ports. No problem. Locks can be built in the barrage. Also, dredging can stop navigation channels from silting up.

▶ *Flood protection.* The risk of sea and river flooding behind the barrage will be less because the high tide will be lower.

▶ *Recreation.* At present the estuary is not used much for recreation because of the high tides, strong currents, and wide mud flats. Angling and birdwatching are important in the upper estuary.

▶ *Water Quality.* If the scouring force of the ebbing water is reduced, pollution from the towns may build up behind the barrage.

▶ *Employment.* Building the barrage will create 30 000 jobs for 10 years but only 500 permanent jobs.

▶ *Fish.* Salmon, eel and sea trout migrate upstream to spawn. They could swim up through the sluices and ship locks.

▶ *Birds.* The Severn estuary is internationally important as a feeding and breeding area for waders. Slimbridge Wildfowl and Wetlands Trust is situated here. A decrease in the tidal range would reduce the feeding area for these birds. Bird populations could fall.

To build or not to build? . . . That is the question

Scientists and engineers completed a £6 million research programme in November 1989 to see if this barrage could be built. It is technically possible though it would be expensive.

The pressure group Friends of the Earth do not like the scheme. They claim it could lead to an environmental disaster. They give two main reasons:–

1 It will change the ecology of the estuary. Internationally important wildfowl habitats will be lost.

2 The money needed to build the barrage would be better spent on energy conservation. For example on combined heat and power, and windpower schemes. They think these would be much more effective in helping to stop the **greenhouse effect**.

QUESTIONS

Copy and complete these sentences using the correct word in the brackets.

1 (a) Tides are caused by the gravitational pull of the (stars, earth's magnetic field, moon).
(b) A tidal barrage harnesses (renewable, nuclear, fossil) energy.
(c) One of Britain's best sites for a tidal power scheme is the (Humber, Thames, Severn) estuary.
(d) This is a good site because the estuary has a very high (rifle, tidal, poultry) range.

2 List three advantages of the scheme.

3 List two problems the scheme might cause. Say if they can be solved.

4 Copy figure 14.2 into your book. Fill in the missing words. Use an atlas to name the towns.

Research

A plan has been drawn up for tidal power on the Mersey estuary. Find out about this plan.

Sitting on a volcano

Figure 15.1   Three uses of geothermal heat

Source	PJ	%
Hydro	40	43
Geothermal	26	28
Oil	23	25
Coal	3	4

Table 15b   Energy use in Iceland

Either you've got it or you haven't

With your partner study Iceland's fact file (table 15a). What kind of place is Iceland? Do you think it has much going for it?

You might now be thinking that Iceland has very little going for it. The country is cold, wet, dark in the winter, has very little good soil and no fossil fuels. But two important words are missing from Iceland's fact file and they make all the difference. The words are **geothermal** and **hydropower**.

No other country in the world is more fortunate than Iceland when it comes to these two renewable energy resources. It has such massive reserves of both that it could be renamed Renewable Energy Island.

Per person, Icelanders are the world's number one user of geothermal energy and number two user of hydroelectricity.

Location	North Atlantic, near Arctic Circle Remote . . . and sitting on a volcano		
Land	total		103 000 km²
	lakes		3000 km²
	sands and lava		15000 km²
	glaciers and wasteland		67 000 km²
Climate		Year	January	July
	temperature (°C)	4.3	−0.9	10.6
	precipitation (mm)	7	68	51
Soil	very little – mostly infertile		
Energy	coal – none, oil – none, gas – none		

Table 15a   Iceland's fact file

Iceland sits astride the Mid Atlantic Ridge where two plates of the earth's crust are slowly drifting apart. As the plates spread, hot **magma** from the earth's mantle rises up into the crust. Any cold water that comes into contact with the hot rock is heated and rises through fissures to the surface as steam or hot springs. The steam can be used to generate electricity, the hot water for district heating.

Uses of geothermal heat

▶ Space heating. 82% of Iceland's homes are heated with geothermal energy. Reykjavik's district heating scheme is the largest in the world. Iceland has 30 geothermal district heating schemes.
▶ Seaweed drying for the pharmaceutical industry.
▶ Manufacturing chemicals.
▶ Fish farms.
▶ Three other uses are shown in figure 15.1.

Sudernes regional geothermal scheme

▶ The geothermal station is sited at Svartsengi.

▶ The geothermal fluid is hot brine.

▶ The hot brine heats cold water in heat exchangers.

▶ The hot water is then piped to towns, villages and Keflavik airport.

▶ The pipelines are insulated with rockwool. They run above ground. Only in large towns are they below ground.

▶ Electricity is also generated and sent to the towns.

Figure 15.3   Svartsengi geothermal station

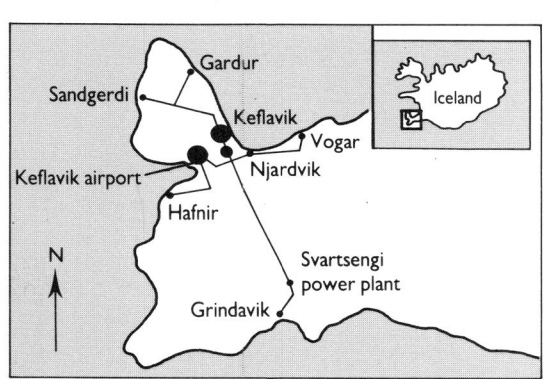

Figure 15.2   Sudernes regional scheme

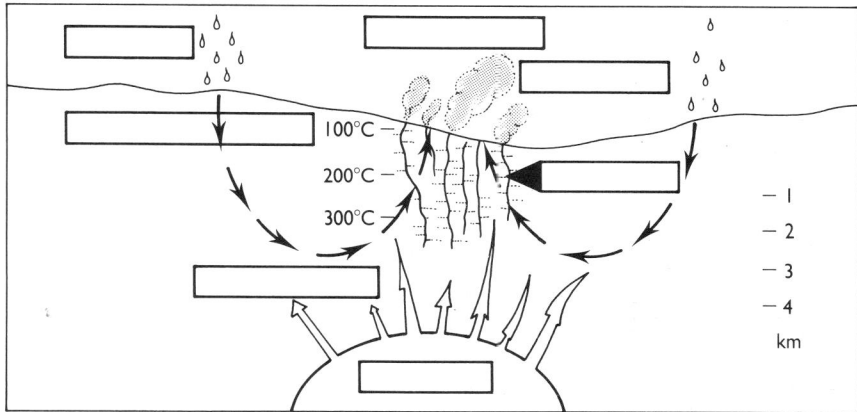

F gure 15.4   A geothermal system

Core   Work in pairs.

1 Copy and complete this paragraph using these words:

- hydropower ● fossil fuels ● − 0.9°C
- geothermal energy ● space heating
- renewable.

Iceland has a cold climate. The average January temperature is _____. About one third of all energy in Iceland is used for _____. The country has no _____   _____. Iceland has massive reserves of two _____ energy resources. These are _____ and _____ _____.

2 Draw and label figure 15.4 using these words:

- rain ● fissures ● hot water ● magma 1200°C ● steam ● cold water ● thermal area.

3 What percentage of Iceland's energy comes from,
(a) fossil fuels   (b) renewable energy.

4 Can you list seven uses of geothermal heat in Iceland?

5 What do you think are the environmental impacts of the Sudernes scheme?

Energy future ... the next hundred years

Your island ... situation 1990

▶ In 1990 coal, oil and gas supplied all your island's energy needs.

▶ Your island's coal reserves are falling. You must conserve coal. In future you will need to convert coal to petrol for the island's transport.

▶ All your oil and gas are imported. World reserves of these fuels are falling fast. Reliable supplies are hard to find. They are also becoming more expensive.

▶ You and three others are the island's energy committee. You must plan your island's energy future.

Rules
Only two. You must;

▶ Supply enough energy to meet your island's energy need at each future date

▶ Protect the environment as far as possible

Getting started

▶ Study the map of your island shown in figure 16.1.

▶ Study table 16a which shows your island's untapped renewable energy resources.

▶ Copy table 16b into your book.

Now it is up to you
Start planning for the next 100 years.

1 In the year 2000 your island will require 100 units of energy. Coal, oil and gas can supply 45 + 35 = 80 units. You must supply 20 energy units from renewables.

2 Study figure 16.1 and table 16a again. Then choose your energy options. You can make up to three choices. Write down the amount of energy your options can supply in the correct boxes in the year 2000 row in table 16b.

3 Add up the energy units in this year 2000 row. Make sure you supply 100 units.

4 *Remember. These options are renewable. They will continue to supply energy units in 2020, 2040, and 2060. So, put the energy units they can supply in the boxes for these years NOW.*

5 Now go on to year 2020, then 2040, then 2060.

6 Lastly. For good energy planners only. Add years 2080 and 2100. You choose the island's energy need. You decide the energy unit for coal. Oil and gas have run out by now.

Can you now meet this new energy need? Do you need to consider new renewable sites? Are there any?

Renewable	Site		Energy units
Hydro	H	1	5
	H	2	10
	H	3	5
Tidal	T	1	15
	T	2	15
	T	3	20
Wave	Wa	1	15
	Wa	2	5
Solar	/		5
Biofuel	B		5
Geothermal	G	1	10
	G	2	5
Wind	W	1	10
	W	2	10
	W	3	10

Table 16a
Your island's renewable energy resources

Figure 16.1   Your island

Year	Energy need	Coal	Oil & gas	Units required	Hydro			Tidal			Wave		Sol	Bio	Geo		Wind			Energy supplied
					H1	H2	H3	T1	T2	T3	Wa1	Wa2			G1	G2	W1	W2	W3	
1990	100	60	40	0																100
2000	100	45	35	20																
2020	100	35	15																	
2040	100	25	5																	
2060	100	15	0																	

Table 16b   Energy future

Figure 17.1   Derwent Reservoir

unit 17

Water water everywhere

There is a lot of water on this planet. In fact 1400 000 000 km^3 of it but 97.3% is in the oceans and another 2.1% is locked in the ice caps. Only 113 km^3 falls as rain on the continents. This small amount sustains all human and natural life on earth. It must be used and managed very carefully.

Local needs

Table 17a shows how the Sunderland and South Shields Water Company has developed four water resources to meet the growing water demand of the towns of Sunderland and South Shields.

Until 1940 the company relied on groundwater supplies. As the demand for water grew two reservoirs were built in the Pennines. When demand increased further, water was taken directly from the River Wear. Water from the Kielder Reservoir is transferred from the Tyne drainage basin to the River Wear. It can also be piped to the Derwent Reservoir.

Water

Year	Ground Water	Burnhope Reservoir	Derwent Reservoir	River Wear	Total Resource	Demand
1900	8.20	–	–	–	8.20	7.70
1920	10.00	–	–	–	10.00	9.70
1940	13.40	–	–	–	13.60	13.60
1954	12.44	5.0	–	–	18.08	17.11
1967	13.74	4.4	10.0	–	28.14	25.49
1978	12.69	4.4	11.8	5.2	34.09	29.50
1982	12.69	4.4	11.8	10.0	39.69	31.50
1988	9.23	4.4	17.0	10.0	40.63	31.85

Table 17a   Local water supply (mg)

National needs

There are ten water authorities in England and Wales. They use surface water and ground water to supply four users.
Table 17b shows,
(a) how each water authority obtains its water.
(b) how much is supplied to each water user.

Water Authority	Resources %		Water Users %			
	Surface Water	Ground Water	Public Supply	Agri-culture	CEGB	Indus-try
Anglian	50	50	84	3	0	13
Northumbrian	89	11	93	0	0	7
North West	84	16	70	0	3	27
Severn-Trent	82	18	40	0	52	8
Southern	26	74	88	1	0	11
South West	85	15	80	5	7	8
Thames	58	42	93	0	3	4
Welsh	99	1	24	0	73	3
Wessex	53	47	87	2	0	11
Yorkshire	89	11	59	1	27	13
England and Wales	72	28	72	1	17	10

Table 17b   Regional water supply and use 1988

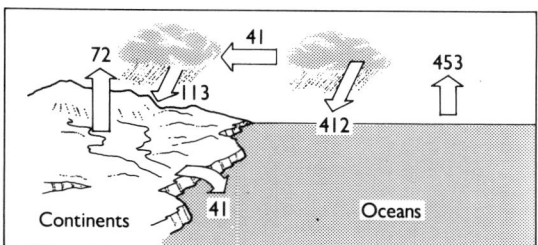

Figure 17.2   Continental rainfall (km^3)

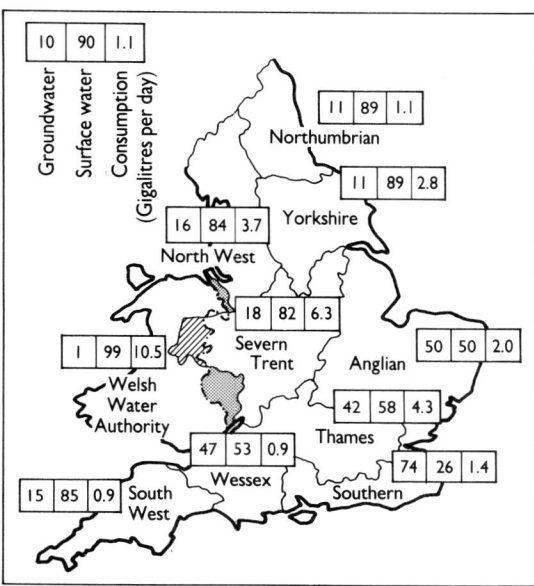

Figure 17.3   National water resources 1988

Global water supply

The biggest problem with the world's water supply is that it is not spread around evenly. Some areas of the world get more water than they need. Other areas do not get enough. Figure 17.4 shows which areas have a surplus and which areas have a deficit.

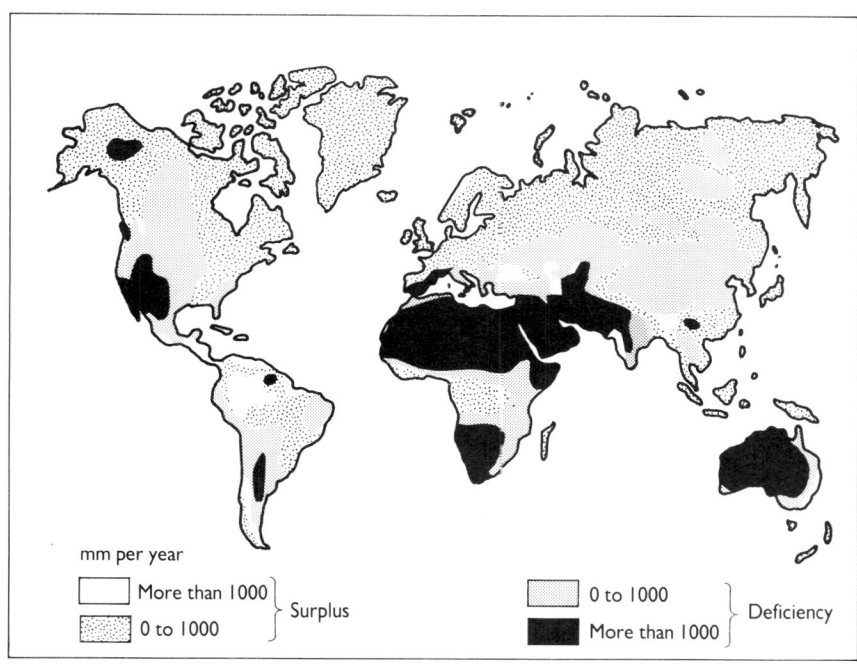

Figure 17.4   Global water surplus and deficit

Q U E S T I O N S

Core

1 Name the two sources of water supply in Britain.

2 Give the four main water users in Britain.

3 Copy and complete.
The global water supply is not distributed evenly. Some areas have a water surplus. This means _____.
Other areas have a water deficit. This means _____.

4 Name two areas in the world with water deficits.

5 Draw the bar graph opposite. Mark it for your water authority and show
  (a) the two sources of water supply
  (b) the water users

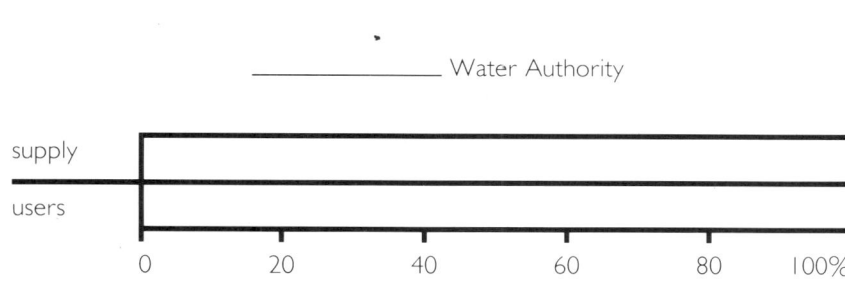

_____ Water Authority

A **1** In what year did the Derwent Reservoir first supply water?

  2 In 1988 which water authority obtained most of its water supply from groundwater?

B **1** List in rank order the amount of water supplied to the Sunderland and South Shields Water Company by groundwater, reservoirs and the River Wear in 1988.

  2 Trace the map of the water authorities. Colour in the groundwater supplies using this key:
  more than 70% – yellow   40–70% – green   less than 40% – blue
  What pattern emerges? Why? You may need to look at a relief and rainfall map of Britain to give you clues.

Dam problems

You can't please everybody

An area of land can be used for several purposes. For example,

- farming
- houses
- factories
- airports
- road
- recreation
- nature reserves
- reservoirs

Different groups of people might want to use the land for different activities. Some of these activities may compete with each other and cause conflict. If the land is used for a reservoir, only two of the other uses in the list above can take place as well. Which ones are they?

Site		Yield (mgd)	Cost (£m)	Disturbance	Completion
Cow Green	1	35	2	SSSI	1969
Harwood	2	31	5	SSSI, farmland, houses, roads	1971
Upper Cow Green	3	30	6	Nature reserve	1971
Lower Maize Beck	4	20	4.5	Nature reserve	1971
Upper Maize Beck	5	20	4	Nature reserve	1971
Middleton	6	75	6	Farmland, roads, houses	1972
Eggleton	7	32	5.3	Road diversion	1971

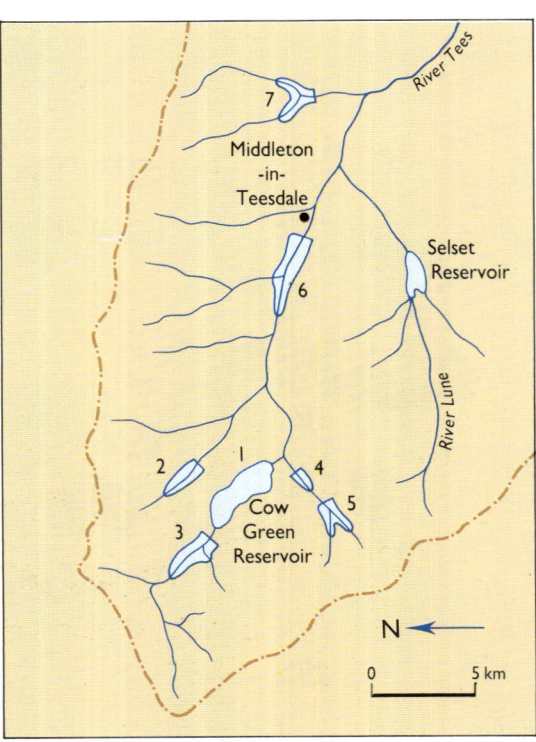

Figure 18.1   Proposed reservoir sites

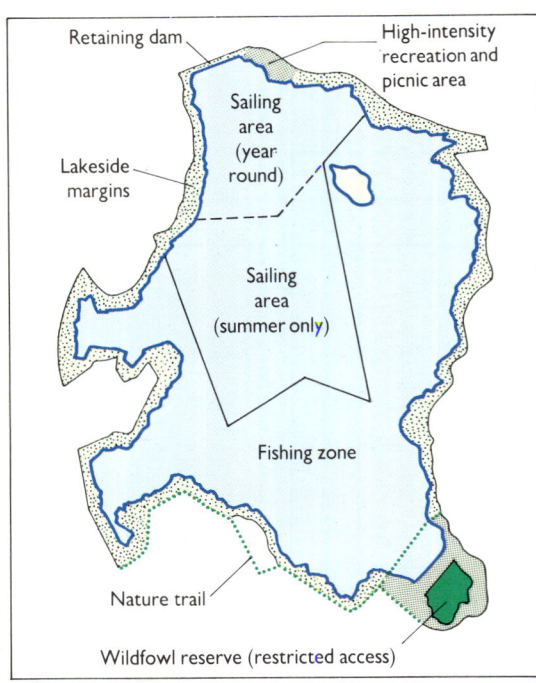

Figure 18.2   Zoning the Chew Valley Reservoir

Case 1

In the 1960s a local water authority planned to build a new reservoir. Where should the reservoir be built? Seven sites were proposed. Each site would cause some conflict with other land users. For each site there were also differences in costs, water storage and the length of time to build the reservoir. Figure 18.1 and table 18a gives this information for the sites.

Eventually, a House of Lords Committee decided that the site causing least overall conflict was the Cow Green site. The water authority built the reservoir there.

Many people disagreed with this decision. Those against it included the Nature Conservancy Council, the National Parks Committee and local conservationists. Their main objection was that the reservoir would flood land occupied by rare alpine flowers.

Case 2

In the 1950s Bristol needed more water. Chew valley was chosen as the site for a reservoir. It had good hydrological conditions and an unpolluted catchment. It was also first class farming land. When the reservoir was built farmers lost valuable land.

Another problem surfaced a few years later. Fertilizer sprayed on surrounding fields drained into the reservoir. Water weed and algae grew and choked the lake. When they died bacteria decomposed them and used up oxygen in the water. Lower oxygen levels brought a decline in the fish population.

A third problem arose in the 1970s. Different groups of people wanted to use the reservoir for different activities. These activities included sailing, fishing, bird-watching, walking and picnicking. The reservoir was zoned to keep conflicting activities apart. Figure 18.2 shows how this was done.

Key:

▬▬▬▬	Dam
● ● ●	Houses Farms
⋎⋏	Quarry
⌒50⌒	Contours
 NR	Nature Reserve
A66	Roads
🌳🌳	Woodland
⊢⊢⊢⊢	Railway

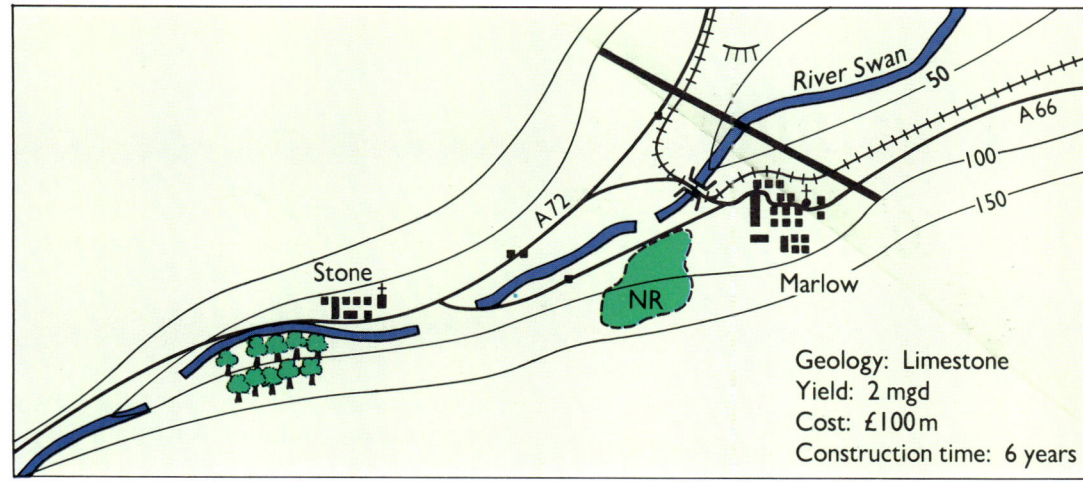

Geology: Limestone
Yield: 2 mgd
Cost: £100m
Construction time: 6 years

Figure 18.3 Swan site

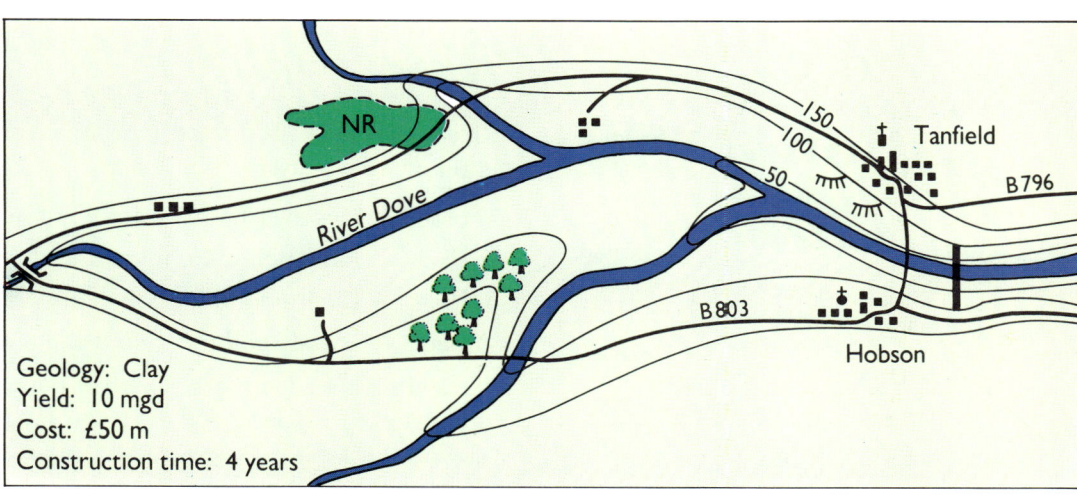

Geology: Clay
Yield: 10 mgd
Cost: £50 m
Construction time: 4 years

Figure 18.4 Dove site

Q U E S T I O N S

Small group work

A nearby town will need 5 million gallons of water a day in 5 years time. Two sites have been proposed for a reservoir. These are shown in figures 18.3 and 18.4. Your group is in charge of appraising the sites.

(a) Carefully study the proposed sites for the new reservoir.
(b) Write a two paragraph letter to the local water authority to say:
   1 Why the Swan site is totally unsuitable.
   2 Why you would choose the Dove site.
(c) You have also been asked to zone the Dove site for several activities. These are:- fishing, sailing, windsurfing, walking, bird-watching, water skiing, a nature reserve, holiday chalets, picnicking and a motor cruiser trip.
   You will also need car parks, toilets, an information centre and two or three shops.
(d) In your group discuss how you are going to zone the reservoir and the surrounding land for these activities.
(e) Draw a large outline of the Dove site on A3 paper. Show how you would zone the reservoir. Provide a suitable key for your map. Mount the map for display.

Core

1 Why was the Cow Green site chosen?
2 Why did some people object to this choice?
3 Describe one problem caused by building the Chew Valley Reservoir.

A warning . . . just to swimmers?

Why does the newspaper cutting advise swimmers not to swim in the River Ouse?

Swimmers warned of pollution threat

Competitors who took part in a river race last Saturday have been warned they could become ill after the River Ouse, where the race took place, was found to be heavily contaminated with sewage.

Levels 280 times higher than European guidelines for bathing beaches were recorded. Swimming in estuaries and rivers is officially discouraged as there are no sewage limits.

Figure 19.1

Water Authority	Classes 1 and 2 %	Classes 3 and 4 %
Anglian	92	8
Northumbrian	98	2
North West	80	20
Severn-Trent	90	10
Southern	94	6
South West	88	12
Thames	95	5
Welsh	93	7
Wessex	95	5
Yorkshire	87	13
England and Wales	91	9

Table 19a   River water quality

River uses

Rivers are important in Britain. They are used for: • water supply   • commercial fishing   • recreation   • wildlife conservation

River pollution

A river becomes polluted when the purity of its water is destroyed. Sometimes the pollution is natural. Most of the time it is caused by people. Pollution needs to be controlled so that the river can be used for the purposes given above.

Table 19b shows the main types of pollution caused by humans, where they come from and what effect they can have on a river.

River quality

Rivers are checked to see if they are clean or polluted. They are put into four river classes. These are:

1 Good – The river has a clean bill of health and lots of fish.

2 Fair – Some pollution but still fairly good for fish.

3 Poor – Much pollution. Few if any fish.

4 Bad – Grossly polluted. No fish.

Table 19a gives some data for river pollution in the ten water authorities in England and Wales in 1988.

Pollutant	Main source	Effect on the river
Sewage	Towns	When bacteria decompose the sewage they use up oxygen in the river. Some fish species need a high level of oxygen in the river. (Sewage also contains nitrogen.)
Chemicals	Factories	Some chemicals are toxic. They poison fish. Chemicals can bioaccumulate and biomagnify in food chains to poison river carnivores.
Hot Water	Power stations	Warm water holds less oxygen than cold water. Some fish species prefer cold water.
Fertilizers	Farms	Nitrogen and phosphorous are plant foods. Washed into rivers by rain they promote the growth of river algae. When bacteria decompose dead algae oxygen is used up. Nitrogen in the public water supply has been identified as a health hazard.
Pesticides	Farms	Can be toxic. Can biomagnify and bioaccumulate in food chains.
Suspended solids	Mines	Dirties the water. Reduces sunlight needed by river plants for photosynthesis.

Table 19b   Pollutants, sources and effects

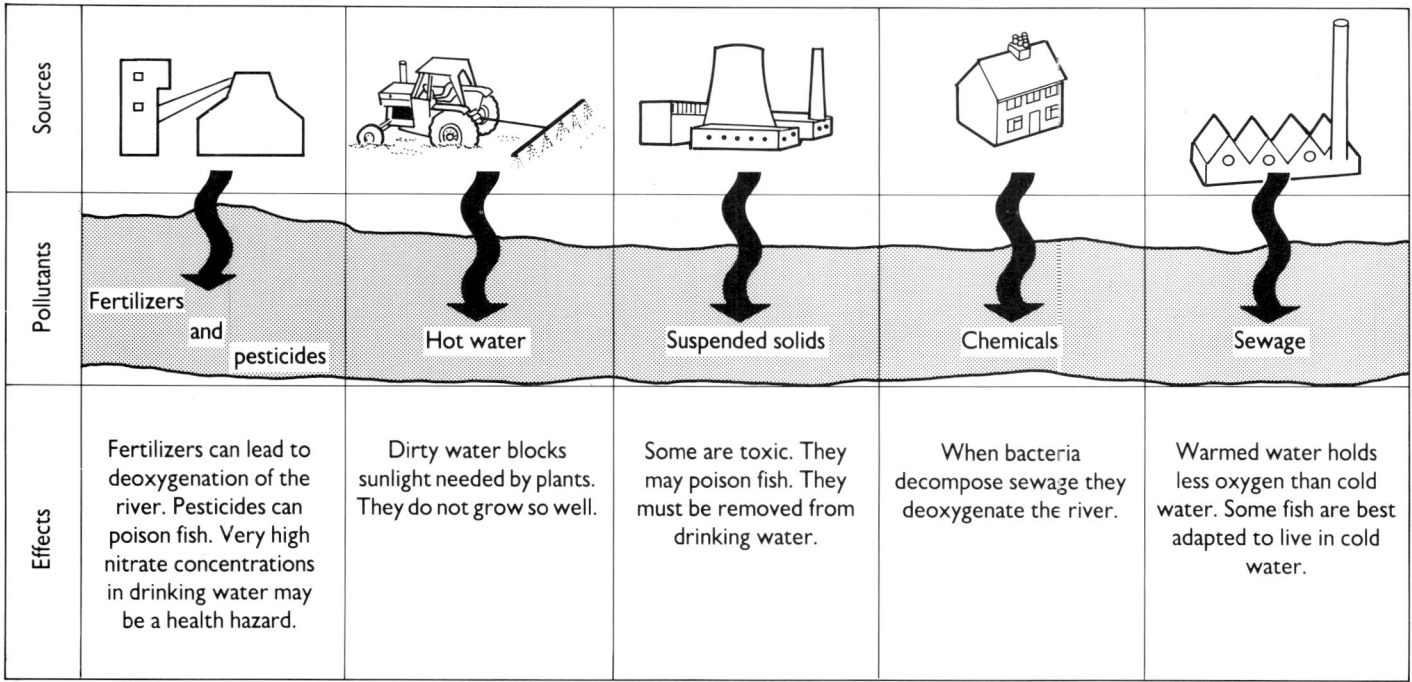

Figure 19.2   River pollution

Year	No.
1980	12500
1981	12600
1982	13100
1983	15400
1984	18600
1985	20000
1986	21400
1987	23300
1988	26900

Table 19c   Pollution incidents

Polluter	Incidents	%
Industry	9085	38
Farms	3952	17
Sewage	4578	19
Other	6039	26

Nothing found in 3272 incidents.

Table 19d   Pollution culprits (1988)

Polluter	Prosecutions	%
Industry	133	41
Farms	188	57
Sewage	6	2

Table 19e   Prosecutions (1988)

Good but can do better

The general condition of our rivers is pretty good but pollution incidents are steadily rising while the actual number of prosecutions is very low.

In 1989 the National Rivers Authority took over the job of policing our rivers from the water authorities. They may be much more strict in dealing with polluters.

Q U E S T I O N S

Core

1 Give four uses of rivers.

2 Name the four river classes.

3 Would fish live in river class 4?

4 Figure 19.2 shows the sources, pollutants and effects involved in river pollution. The effects are in the correct order but the other diagrams have been mixed up. Match the sources and pollutants correctly.

A Study tables 19a, c, d and e.
  1 What percentage of your water authority's rivers are:
     (a) Polluted   (b) Fairly clean?

2 Did the number of pollution incidents rise or fall between 1980–88?

3 Who is the biggest of the four water polluters?

4 Which polluter is least responsible for pollution but is most often prosecuted?

B Study tables 19a, c, d and e.
  1 Name the three water authorities with the highest percentage of polluted rivers.

2 Between which two years was the greatest increase in pollution incidents?

3 What percentage of pollution incidents in 1938 were prosecuted?

Does it still deserve the title?

Europe's open sewer

The River Rhine was given this title in the 1950s and 1960s. It deserved it because it was so badly polluted.

When a river is polluted it can recover naturally but natural recovery is a slow process. It can be several miles downstream from a major point of pollution before the river becomes clean again. If more pollution enters the river before it can recover it has no chance. A bit like a boxer being knocked down for the sixth time in the same round. Can you suggest from figure 20.1 why the River Rhine did not have much of a chance?

The Rhine is vulnerable because it flows through some of Europe's largest industrial areas. Each area dumped its own pollution into the Rhine before the river could recover from the previous dose of pollution.

Pollutants

▶ *Chemicals*   Waste chemicals have been dumped into the river for years. Several of Europe's largest chemical factories are located on the Rhine and its tributaries. Some chemicals are very toxic to river life. They are also a problem if water is to be extracted for drinking.
Salt from French potash mines in Alsace has been dumped into the river.
In November 1987 a chemical spillage from a Swiss factory in Basle caused very serious pollution of the Rhine. The West German government demanded that the chemical company pay compensation.

▶ *Sewage*   Many towns use the Rhine to dispose of their sewage. In the past sewage was untreated or partly treated. Bacteria oxidised the sewage using some of the river's oxygen. This lowered the amount of oxygen in the river. The river became deoxygenated. People also used the sewage system to get rid of other waste from their houses. Pills, paint, used motor oil and batteries were flushed down toilets and poured down drains.

▶ *Nitrogen*   Fertilizers are spread on fields in the Rhine drainage basin. Crop roots take up some of the chemicals but much is washed into the river by the rain.

▶ *Thermal*   Power stations and some factories pipe warm water into the river. Warmer water holds less dissolved oxygen than cold water. Thermal pollution causes a local problem for several fish species.

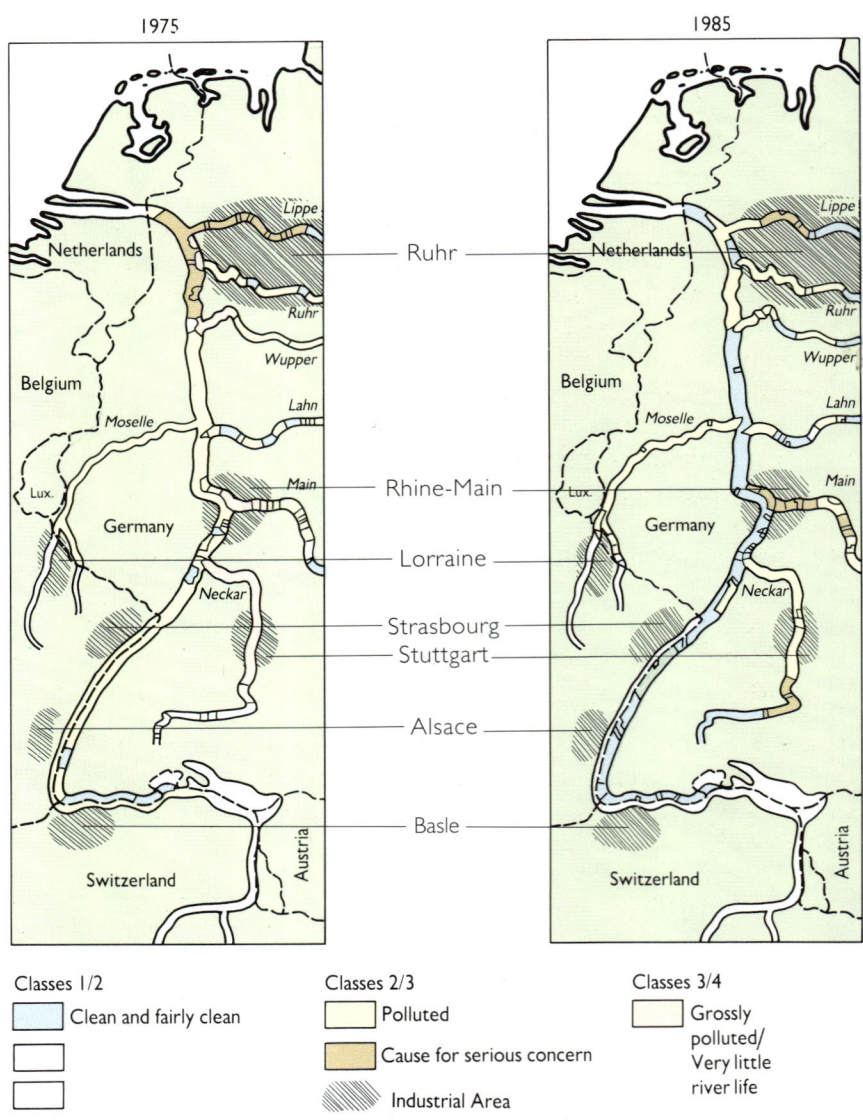

Figure 20.1   River classes of the Rhine

Figure 20.2 Industrial area on the Rhine

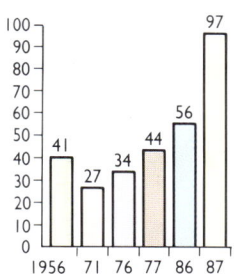

Figure 20.3 Changes in numbers of Rhine species.

Action Since the River Rhine could not recover from pollution by itself it had to be helped.

▶ *Local action* Local authorities have built better sewage treatment works. Some chemical firms now treat their waste chemicals becore piping them into the river or they safely dispose of them elsewhere. Individuals have been encouraged to stop using the sewage system to get rid of pills, paint, motor oil and other household waste. The temperature of waste water entering the river from power stations has been lowered.

▶ *International action* Local action alone could not save the Rhine. In 1963 the Rhine Protection Committee was set up. Member countries included Belgium, France, West Germany, Netherlands and Switzerland. They signed a treaty to:
  ● limit chemical discharges
  ● ban very dangerous chemicals
  ● carry out research on improving water quality
  International cooperation is not easy. France continued to dump salt into the river because other disposal methods were more costly and more technically difficult. The cost for safer disposal was to be split between France, West Germany, the Netherlands and Switzerland.
  The River Rhine is not as important to France as it is to Germany and the Netherlands. In the flat Netherland landscape the Rhine provides about half of the country's drinking water.

Has it worked? Action to save the Rhine has been taken. Has it worked? Has the condition of the river improved? Perhaps the Rhine no longer deserves its old title? You be the judge. Figures 20.1, 20.3 and 20.4 will help you to make a decision.

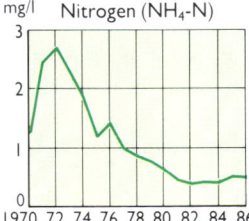
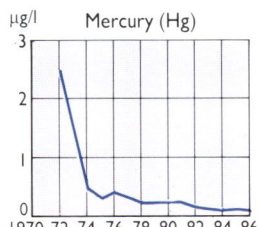
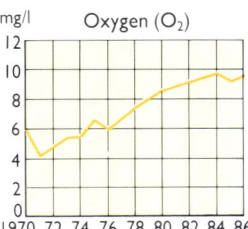

Figure 20.4 The chemical condition of the Rhine

QUESTIONS

Core

1 List the counties which the River Rhine flows through.

2 Name four industrial areas on the Rhine or its tributaries.

3 Name four of the Rhine's pollutants.

4 Describe the action taken to protect the Rhine from pollution.

A 1 Did the oxygen level in the Rhine increase or decrease between 1970 to 1986?

2 In which years were (a) the lowest (b) the highest number of species found in the Rhine?

3 Complete these two sentences.
In 1975 most of the River Rhine was in class ___
In 1985 much of the river was in class _____

B 1 Copy and complete the information below.

	1972	1986	
nitrogen	2.7		mg/l
mercury		0.1	ug/l
oxygen	5		mg/l

2 Describe the improvement in the number of species recorded in the Rhine between 1971 and 1987.

3 Write a short report to describe the condition of the Rhine and its tributaries in 1975 and 1985.

Research

Pollution of the North Sea has been an environmental issue since the early 1980s. In groups of four find out:
(a) the causes of North Sea pollution.
(b) if pollution is being treated.
(c) if any international cooperation is taking place to reduce the pollution.

Safe to swim?

Figure 21.1   Have things really changed?

Progress?

Look at figure 21.1. In Henry's time Queen Elizabeth was on the throne. People had little technology. No cars, no planes, nobody could fly to the moon. That is how they got rid of their sewage.

Four hundred years later Henry's great, great, great, great, great . . . grandson, Fred has a car, a microwave and has seen men land on the moon. Much sewage is still disposed of in a similar way.

How much?

- ▶ 10% of Britain's sewage goes into the sea.
- ▶ This is 300 million gallons every day.
- ▶ 27% of sewage outfalls discharge untreated sewage.
- ▶ 51% of outfalls discharge sewage with little treatment.
- ▶ 69% of discharges are at or above the low water mark.

Tourism

Holiday resorts are big business in Britain. Millions of people are attracted to them each year by their sand and sea. Some people will be at risk. Not so much from Great White Shark attacks but from catching a disease by swimming in water polluted by sewage. The main environmental impacts of piping sewage into the sea are:–

- ▶ *Health*   Sewage contains viruses and bacteria which can cause ear, nose and throat infections, skin irritations and gastroenteritis.
- ▶ *Visual*   Well, would you swim in it?
- ▶ *Wildlife*   Sewage contains nitrogen and phosphorous. These chemicals are fertilizers for seaweed and plankton. These plants grow, block estuaries and block out sunlight needed by other life in the sea. When they die and decompose oxygen in the water is used up.

EC rules

In 1976 the EC set cleanliness standards for the water people swim in. Member countries were asked to;

- Name their popular beaches.
- Check them for pollution.
- Clean up the beaches that failed the test.

Figure 21.2   Fancy a swim?

In 1986 Britain listed 392 beaches. When they were checked 40% of them failed to meet EC standards. In 1987 45% of the beaches failed. In 1988 33% failed. In 1989 440 beaches were tested and 304 passed. What percentage failed?

What is the risk?

Research in the USA found that 18 in every 1000 swimmers became ill after swimming in the sea. Britain's waters may be more polluted. It has been estimated that 31 in every 1000 swimmers could become ill.

Pass or fail

Table 21a shows the results for some of our main holiday beaches from 1986–1989. P is a pass, F is a fail. A beach which passes is given the 'Blue Flag'.

Do you think a beach should be awarded the 'Blue Flag' only if it passes the test every year? Or perhaps three out of four years?

The map shown in figure 21.3 uses a pass rate of three out of four years.

Beach	1986	1987	1988	1989
Bridlington North	P	P	P	P
Bridlington South	P	P	P	P
Scarborough North	P	F	P	F
Scarborough South	F	F	P	P
Southend Thorpe	F	F	F	P
Southend Westcliff	P	F	F	P
Margate Fulsam	F	F	P	P
Ryde	F	F	F	F
Weymouth Central	P	P	P	P
Torquay Oddicombe	P	P	P	P
Penzance Mounts Bay	F	F	F	F
Newquay Fistral	P	P	P	P
Tenby South	F	P	P	P
Southport	F	F	F	F
Blackpool Central	F	F	F	F

Table 21a   Bathing water pollution results

Solutions

1 Building longer sewage outfall pipes to discharge the sewage further out to sea where it will be broken down.

2 Building new treatment works to treat the sewage before it is piped into the sea.

3 Sewage sludge could be dumped at inland sites if any are available.

All three of these measures are likely to increase the domestic water bill.

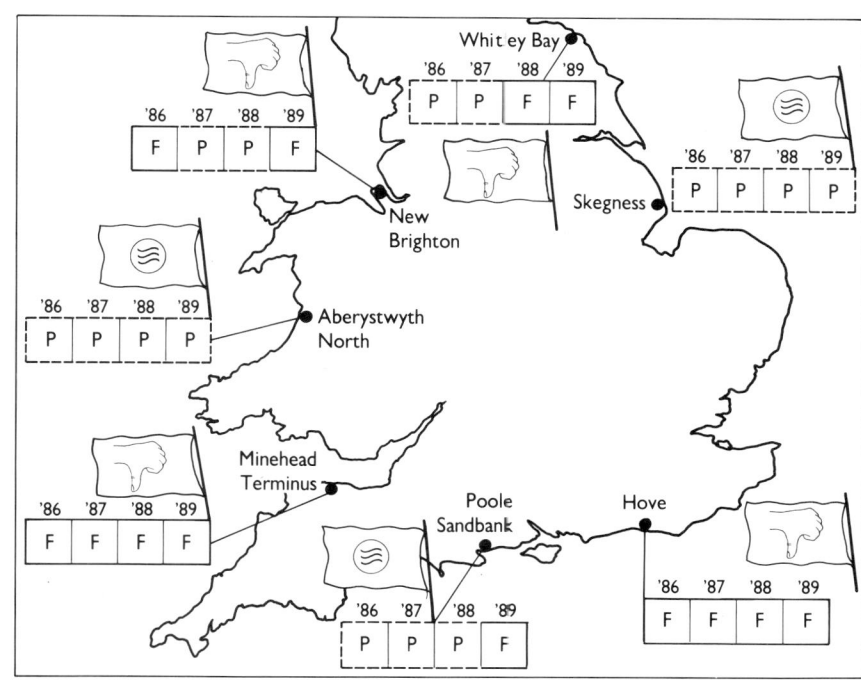

Figure 21.3   Beach awards

Core

1 Copy and complete:
A sewage o_tf_ll pipe discharges sewage into the _e_.

2 List the environmental problems that may be caused by piping sewage into the sea.

3 Suggest two ways to reduce sewage pollution of bathing beaches.

4 Trace figure 21.1 into your book.

A Study table 21a.
1 Did Margate's beach pass in 1987?
2 How many years did Weymouth's beach pass the test?
3 Name a beach which (a) Passed the test (b) Failed the test every year.

B Study table 21a.
1 Which is the best beach in Southend?
2 How many beaches passed every year? Name two of them.
3 Name the beaches with the worst record.
4 Decide on a pass rate for the 1986–1989 results. You can choose three out of four or even four out of four if you wish. Now draw a map like figure 21.3. Plot on your map the results for, Scarborough North Bay, Margate, Torquay Oddicombe and four other beaches of your choice. Award 'Blue Flags' or 'Thumbs Down' symbols.

Wood

unit 22

Wood strategy options

Winter. A cold walk home from school. A hot meal on the table. Gas central heating warms the room. Electricity cooks the meal. Energy on tap. At the press of a switch.

It might be like that in British homes but figure 22.1 shows it is a lot different for some people. In fact, a lot different for a lot of people. Millions of them. Almost every day energy has to be collected and carried to millions of homes in developing countries around the world. No switches either.

In the 1980s two billion people still relied on wood as their main fuel. Wood is a very important energy resource in 95 developing countries. In 21 of them biofuels (wood, charcoal and farm waste) have supplied more than 75% of their total energy use. Table 22a gives some examples. Most wood in these countries is used for heating and cooking in homes. It is also a very important fuel for industry.

Country	%
Burkino Faso	94
India	42
Kenya	74
Malawi	94
Mali	93
Mozambique	89
Sri Lanka	74
Tanzania	89
Zimbabwe	50

Table 22a   Energy supplied by biofuels

The daily drudge

Figure 22.1

Wood in Kenya's energy balance

Figure 22.2 shows the importance of biofuels in Kenya. Kenya's wood resources fall into two groups.

▶ *Forests*   These are generally in remoter areas. They supply about 10% of the country's woodfuel.

▶ *Free-standing trees and bushes.*   These grow individually and in small stands in agricultural land. They supply most fuelwood. Dead branches and twigs are collected for local use by villagers. Dead wood is preferred because it
- is lighter to carry
- is easier to cut
- burns better than green wood

In the villages woodcollecting is normally a woman's job. Wood is also cut and collected commercially and transported to towns. An urban household can spend up to 30% of its annual income on fuelwood. Kenya's total wood stocks have been estimated at about 900 million tonnes.

Future wood use

Kenya's population has one of the highest growth rates in the world. The population could double in less than thirty years. More energy will be needed in the future. The country has no coal. Imported oil could cost more in the future as world oil reserves fall. Nuclear and hydroelectric power are very expensive energy technologies. Kenya is not a rich country. It may have no alternative but to rely on wood as the main fuel.

Areas of desolation are growing around towns as the wood is used up. Trees in the agricultural land and even in the forests are under greater pressure. Soil erosion occurs when trees are felled. The wood resource is beginning to fail. Table 22b shows Kenya's past wood use and projections for wood use up to the year 2000.

Year	Wood use (million tonnes)
1980	20
1985	26
1990	32
1995	40
2000	49

Table 22b   Projected wood use in Kenya

A team of Swedish scientists have done some research for the Kenyan government. Their plan to give Kenya a sustainable supply of wood is shown in table 22c. This plan cannot meet future demand for wood alone. Other things will also have to be done. For example,

▶ burn wood more efficiently in new stoves.

▶ use low cost solar power and wind power.

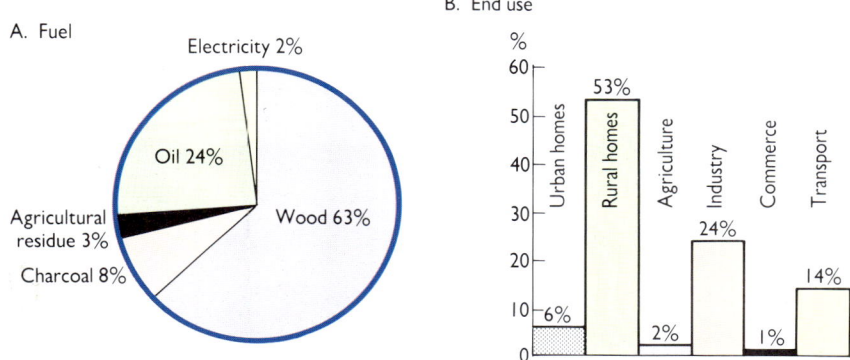

Figure 22.2

Core

1 Copy and make the correct choice from the brackets.
 Many (developed, developing) countries rely on wood as their main fuel. (African, European) countries especially rely on wood. For example, biofuels supplied (74%, 94%) of (Malawi's, Zimbabwe's) total energy in the 1980s.

2 Use table 22a to make a rank order list of countries where biofuels supplied more than 70% of total energy.

3 Which end use consumes most wood in Kenya?

4 Use table 22b to draw a histogram to show Kenya's increasing use of wood.

5 How much wood can the sustainable plan yield every year? Will this supply all of Kenya's needs in 2000?

6 Suggest a renewable energy resource that Kenya could develop.

Option	Description	Sustainable yield (million tonnes)
1 Agroforestry	Grow trees on farmland as an energy crop.	10
2 Replant forest	Clear fell sections of forest then replant	6
3 Periurban plantations	Plant trees around towns	3
4 Industrial plantations	Change of some land from agriculture to fuel plantations	1
5 Natural forests	Manage for fuel – thin, clearfell, replant	2

Table 22c   Wood strategy options for Kenya

Wanted – global solution for a global problem

Secrets and treasures

The Indian and the frog shown below live in a special **ecosystem**. It has been called a living laboratory that has given up only a fraction of its secrets and treasures. Figure 23.1 shows some of these 'treasures' or environmental services.

GENE POOL
Rainforest plants Provide medicines, fibres, resin, rubber, timber and other materials.
Animals Provide food for natives. Carry immunities against pests.
Insects May be used for biological control of pests.

Climate The forests mop up CO_2 during photosynthesis. This helps reduce the greenhouse effect. They also produce massive amounts of oxygen and help control the local climate.

Watershed management The forests control water flow between the atmosphere and the ocean. This helps reduce soil erosion and flooding.
Habitat They are home for thousands of Indians and more than 50% of all species of life on earth.

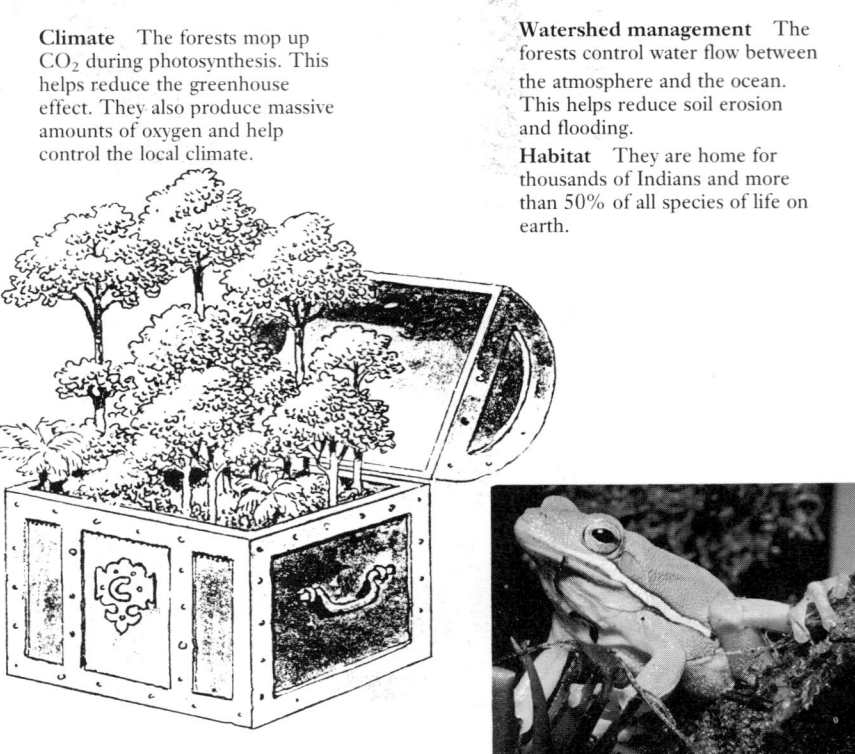

Figure 23.1 Secrets and treasures of the rainforest

What is it . . . where is it?
This ecosystem is the tropical rainforest. Figure 23.2 shows that these forests form a ring around the earth's equator.

Now you see it . . . now you don't
Rainforests have become a global issue because they are being destroyed. They

▶ once covered 16% of the earth's land surface now cover 7%

▶ have shrunk from 24 to 9 million km²

▶ are being destroyed at 200 000 km² a year or 100 acres a minute

▶ will all be gone in 170 years if the present destruction rate continues

▶ could be gone from some countries in just 25 years

The main causes of destruction are shown in table 23a.

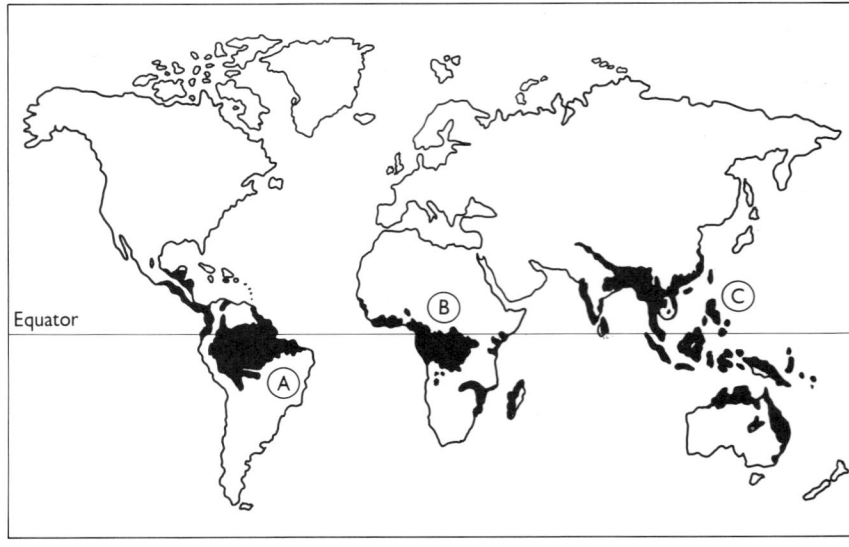

Figure 23.2 Tropical rainforests

The pattern of destruction

In Brazil and central America the pattern of destruction has often been,

Stage 1   Loggers
- build roads
- remove trees
- then move on.

Stage 2   Landless poor
- move into forests along loggers' roads
- burn the forest, grow crops for three years
- move on when the soil loses its fertility.

Stage 3   Cattle ranchers
- move in after the landless poor
- raise beef cattle for about five years
- abandon ranches when the soil loses its fertility.

	%
Commercial logging	17
Fuelwood	10
Cattle ranching	7
Slash and burn farming	62
Plantations	2
Other (roads, dams, mines)	2

Table 23a   Causes of destruction

Core

1 Why have rainforests become a global issue?

2 List the six causes of rainforest destruction in rank order.

3 Copy the map 23.2. Name the three rainforest areas marked A, B and C.

4 Complete this diagram to show the pattern of destruction.

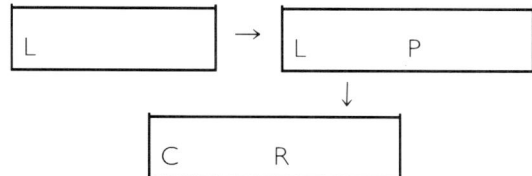

5 Explain why rainforests are a global issue.

Why a global issue?

Rainforests have become a global issue because three main groups of people have conflicting views on the way they should be used. In Brazil these conflicts are;

Indians
- They live in the Amazonian forests.
- In 1500 8 million lived in Amazonia.
- Now there are 250 000.
- Their lands have been taken for farms, hydro schemes, plantations, mines, timber and cattle ranches.
- Some have even been killed.
- They want to keep their homelands and way of life.

Brazilian government
- Sees its rainforests as a valuable resource which it can use to improve the living standards of the people.
- Rainforests can supply timber and minerals to earn foreign money.
- Rainforests can supply land for farms for poor people.
- Brazil needs electricity to develop. The choices for producing electricity are (a) burn coal   (b) expand nuclear power (c) build hydro schemes on rainforest rivers.

Rich countries
- Are worried that destroying rainforests will add to the greenhouse effect.
- Are concerned that the valuable gene pool which they use will be lost.
- Want to protect the Indians.

Rainforests . . . at the cutting edge of environment versus development

Solutions to this global issue are needed. Few are in sight. One solution could be to encourage sustainable development. Recent research has shown that an owner can make,

- $1000 a year by clear felling the trees
- $7000 a year by selected logging + growing fruit and vegetables.

Class discussion

Discuss these points.

1 Brazil did not protest about how European countries developed their own land. Or about how the Indians in America had their land taken. Has Brazil the right to develop its own land in the way it wishes?

2 Burning fossil fuels in power stations and cars adds to the greenhouse effect more than burning rainforests. Should rich countries be concerned more about energy conservation and reducing the number of cars in their own countries than about how Brazil develops its rainforests?

3 If rainforests are to be protected should rich countries help pay for this protection?

Soil

unit 24

Down to earth

Soil is one of the world's most precious resources. In it we grow,

► crops to eat

► crops to feed livestock

► crops to produce fibres for clothes

It also helps form our landscape and sustains wildlife.

Soil takes a very long time to form. Perhaps thousands of years. It is not an unlimited resource. Worldwide it is a resource which is often abused. In recent years it has been estimated that the world rate of:

► soil formation is 5 tonnes per acre per year.

► soil loss is 8 tonnes per acre per year.

Soil erosion in Britain

This is a recent problem. The two main causes are water erosion and wind erosion.

Figure 24.1   Water erosion

Water erosion
Running water washes soil down slopes. At first this creates small rills. These can deepen to form large gulleys like those shown in figure 24.1.

Wind erosion
Strong winds can blow soil hundreds of metres across the countryside. Some farmers rip out hedgerows to make bigger fields so that larger, more economic machinery can be used. Hedgerows act as wind breaks. When they are removed soil is simply blown away.

East Anglia has been particularly badly affected by this type of soil erosion. Soil which should be supporting crops now lies at the bottom of the North Sea.

Figure 24.2   Wind erosion

Reducing soil erosion

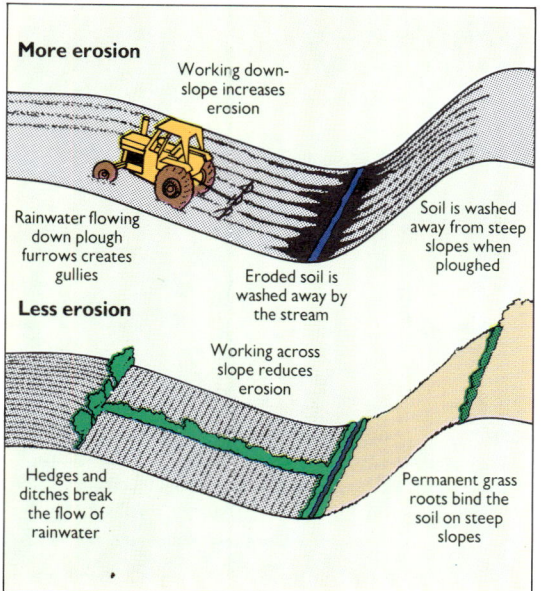

Figure 24.3   Soil erosion

In eastern England marl is mixed with soil. Soil particles stick together. Heavier particles are less likely to be blown away. Farmers can also reduce soil erosion by adopting good farming practices like those in figure 24.3.

Soil erosion – global

Figure 24.4   The Sahel

Figure 24.6   Tropical terraces help reduce soil erosion

Rainbombs

Rainfall is very heavy in tropical rainforests. In the natural forest raindrops hit tree leaves and break up into smaller droplets. The rain eventually reaches the ground as a fine spray. It soaks into the ground and little erosion takes place. When trees are felled raindrops become rainbombs. They explode on impact. Soil particles are hurled into the air and roll down slopes into rivers. Rivers transport the soil to the sea.

Overgrazing

The Sahel area of Africa has often been in the news since 1970. The reason – famine. The cause – drought and soil loss. The Sahel is a fragile area. It has a natural carrying capacity to support a certain number of grazing animals. Digging wells and vaccination against disease has allowed herders to increase the number of animals. The natural carrying capacity is being overloaded.

QUESTIONS

Core

1 Why is soil a precious resource?

2 Complete these two sentences.
   Global soil formation =   ?   tonnes per acre.
   Global soil loss   =   ?   tonnes per acre.

3 Name the two main causes of soil loss in Britain.

4 From figure 24.3 give three ways to reduce soil erosion.

5 Draw figure 24.5. Add these labels in the correct places.
   - rainbombs ● heavy raindrops ● soaks into ground
   - transported by rivers to the sea ● soil washed down slopes
   - explosion on impact ● fine spray

Figure 24.5   Showers or rainbombs

A I Explain how soil is being eroded, (a) n figure 24.1.   (b) in figure 24.2.

B I How is the problem of soil erosion being tackled in the tropical area in figure 24.6.

A solution to pests?

A scientific business

Farming today can be a very scientific business. Many modern farms use technology and materials that were unheard of one hundred years ago. One of the main innovations in modern farming is the use of chemicals. Chemicals are used as,

▶ *Fertilizers* To help plants grow well.

▶ *Pesticides* To reduce damage to crops caused by insects, weeds and diseases.

A growing population needs more food. Table 25a shows the amount of damage done to crops by pests around the world.

The modern pesticide programme began in the 1930s. Since then three generations of pesticides have been developed. All have had some impact on the environment.

First generation

Dichlorodiphenyltrichloroethane (DDT) was the first modern pesticide. It was very effective at killing insect pests. However, it caused two problems because it did not break down quickly but remained in the environment for a long time. Firstly, it killed some harmless insects and organisms in the soil which were beneficial to farming. Secondly, it could build up and concentrate along wildlife foodchains. These processes are called **bioaccumulation** and **biomagnification**. A small amount of insecticide in an aphid could end up being a lot of pesticide in a hawk:

pesticide

5 units		*aphid*
		↓
500 units		*lacewing*
		↓
5000 units		*bluetit*
		↓
25 000 units		*sparrowhawk*

DDT is now banned or strictly controlled in many countries because of its harmful effect on the environment.

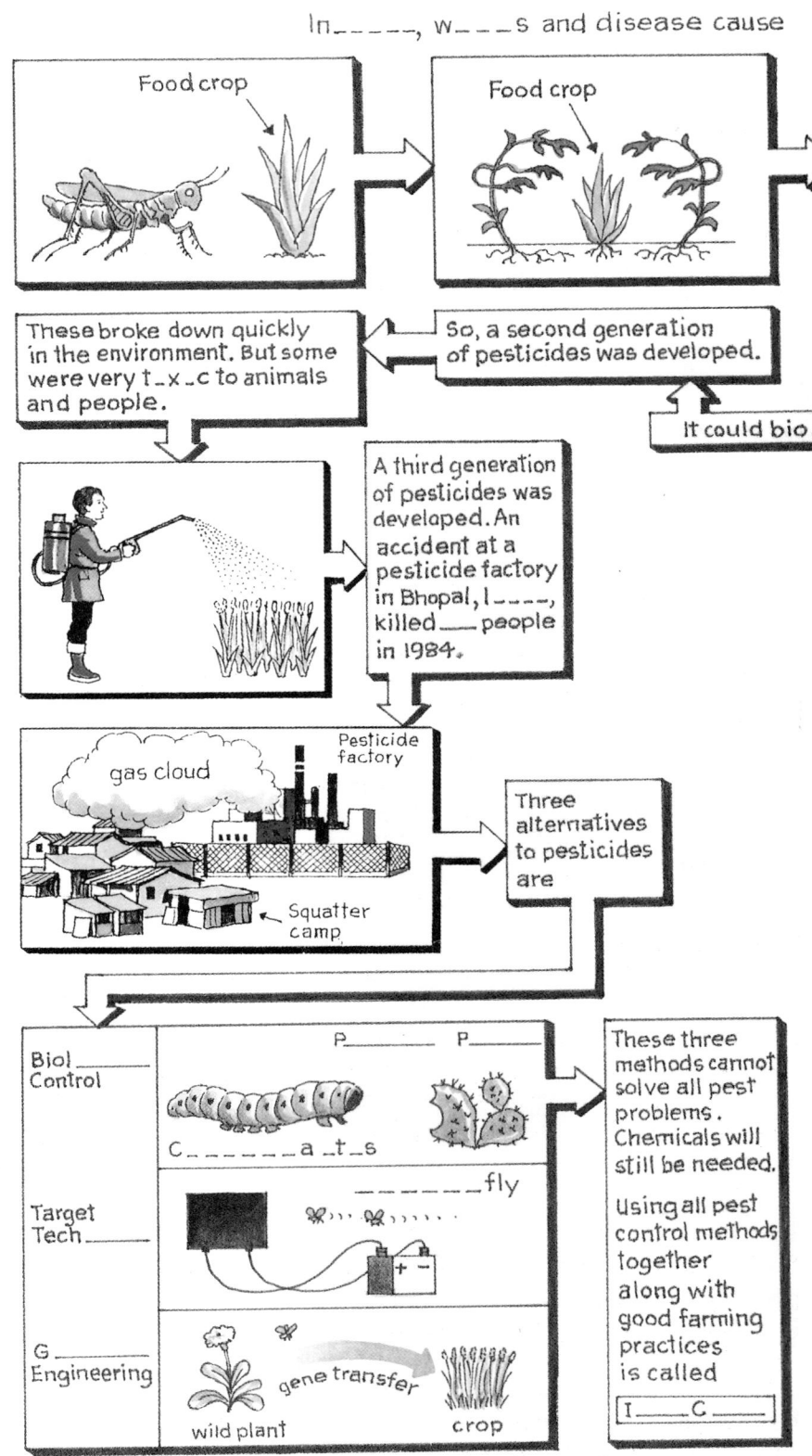

In _ _ _ _ _, w _ _ _ s and disease cause

Food crop

Food crop

These broke down quickly in the environment. But some were very t_x_c to animals and people.

So, a second generation of pesticides was developed.

It could bio_ _

A third generation of pesticides was developed. An accident at a pesticide factory in Bhopal, I_ _ _ _, killed _ _ people in 1984.

gas cloud

Pesticide factory

Squatter camp

Three alternatives to pesticides are

Biol_ _ _ Control

C _ _ _ _ _ a _ t _ s

P _ _   P _ _

Target Tech_ _

_ _ _ _ _ _ fly

G _ _ _ _ Engineering

gene transfer

wild plant

crop

These three methods cannot solve all pest problems. Chemicals will still be needed.

Using all pest control methods together along with good farming practices is called

I _ _ _ C _ _

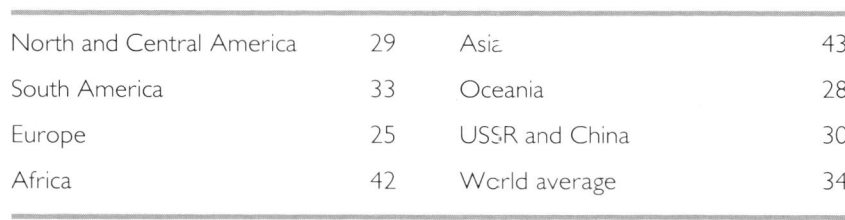

North and Central America	29	Asia	43
South America	33	Oceania	28
Europe	25	USSR and China	30
Africa	42	World average	34

Table 25a   Annual loss to crops (%)

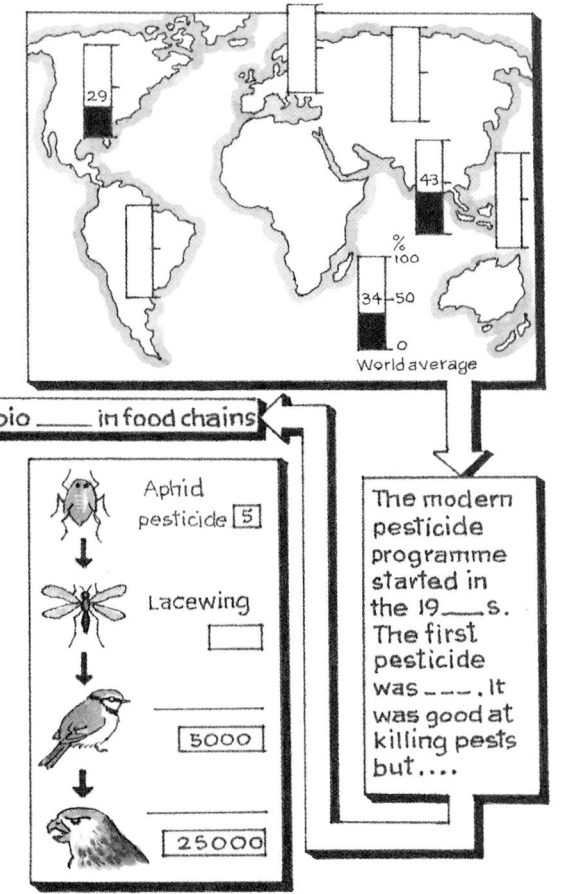

Figure 25.   The pesticide story

Second generation

A second generation of pesticides broke down quickly in the soil and did not bioaccumulate in foodchains. However, some were very toxic to animals and man. Farmworkers in some developing countries have been poisoned because they did not handle these dangerous chemicals properly.

Third generation

Carbamates are a more recent type of pesticide. An accident at a pesticide factory in Bhopal, India in 1984 producing this type of pesticide killed 3000 local people.

Every few years the pest population becomes immune to the pesticide so chemists must manufacture a new pesticide.

Alternatives?

Are there any alternatives to spraying fields with pesticides? Here are three cases.

Case 1   Biological Control
This method uses another organism to control the pest. Earlier this century the 'prickly pear' cactus infested grazing land in Australia. Millions of caterpillars were bred and released. In seven years they munched their way through the invading cactus. These biological controllers were **cactoblastis** caterpillars.

Case 2   Target Technology
Don't go to the pest. Bring the pest to you. In Zimbabwe cattle are affected by the tsetse fly. Research showed that **octonol** and **acetone** in the cow's breath attracted the flies. A decoy was designed and these chemicals were wafted by the wind to tsetse breeding areas. Tsetse are also attracted to black squares. When they flew upwind to the decoy they were electrocuted. This method may become more widely used.

Case 3   Genetic Engineering
Find a plant that a pest does not like. Find the **gene** in that plant which produces the taste that the pest does not like. Transfer the gene to a crop plant. In theory the crop plant should now have the taste that the pest does not like. The pest should leave the crop plant alone. There are great hopes for this method in the future.

Integrated Control

It is no use pretending that biological control, target technology and genetic engineering can solve all pest problems. They cannot. Chemicals will still be needed. However, all of these methods should be used together along with good farming practices wherever possible. This approach to pests is called 'integrated control'.

Work in pairs to complete photocopies of figure 25. 'The pesticide story'. You will need to finish some sketches, complete the map and fill in missing words and numbers.

Limestone quarry near Buxton

Minerals

unit 26

Holes in a National Park

To start this unit copy and complete the four notes below. Work out the answers from figure 26.1.

Notes

1 <u>Uses of limestone</u>   Limestone is used in the _____ and _____ industries and as aggregate to make _____ and to make _____.

2 It is *5 24 3 1 22 1 20 5 4* from *17 21 1 18 18 9 5 19* in limestone rock.

3 <u>Environmental impacts of quarrying limestone</u>
 (a) visual _____-sore on the landscape.
 (b) _____ caused by blasting and lorries transporting limestone.
 (c) _____ caused by blasting and lorries.
 (d) _____ problems.
 (e) loss of _____ land.
 (f) loss of wildlife _____.

4 <u>Peak District National Park</u>
 (a) Lies at the southern edge of the _____ hills.
 (b) Became Britain's _____ National Park in 1951.
 (c) Aims to protect landscape of _____ scenic value.

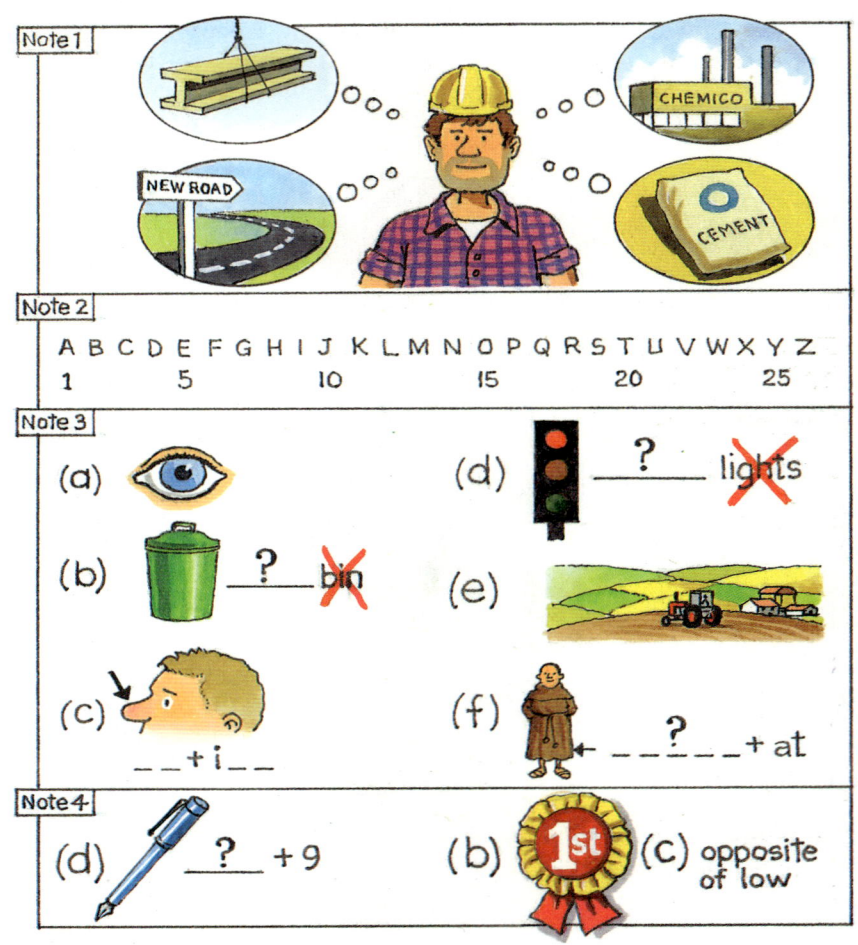

Figure 26.1   Clues chart

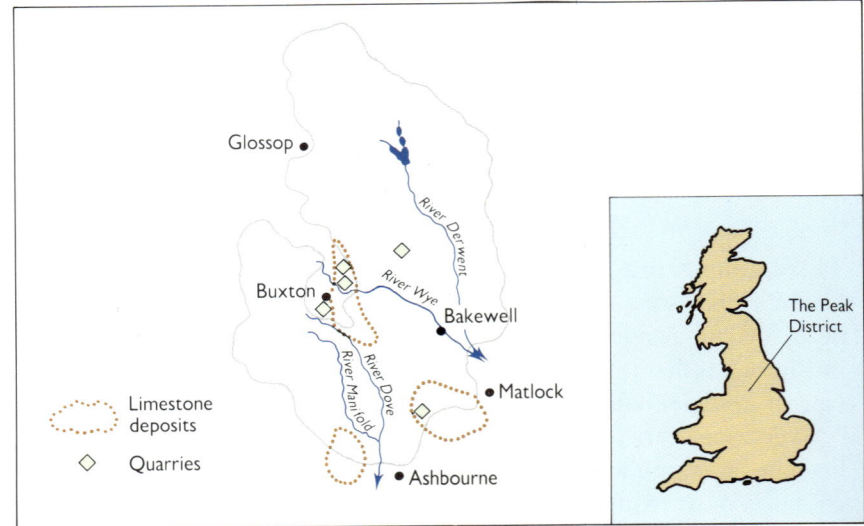

Figure 26.2   Peak District National Park

Figure 26.3   Wolfscote Dale in the Peak District

Figure 26.4   Helping nature

Figure 26.5   Peak District landscape

Conflict in the Peak District National Park

Industrial development and building are restricted in all National Parks to protect the landscape. Most land inside a National Park is privately owned. Although tourism and reservoir building have caused conflict in the Peak District, limestone quarrying has probably caused even greater controversy.

The largest limestone quarries lie just outside the boundary of the Park. Some quarries are worked inside the Park. As the larger quarries are worked out pressure is growing to open up new quarries inside the Park to exploit the limestone reserves there.

Some people argue that new quarries inside the Peak District should be allowed. They say that,

(a) If we need limestone then we must quarry it where we find it.
(b) Limestone quarries provide alternative jobs to farming in this rural area.
(c) The Peak District protects hill and moorland landscape. So do eight of the other National Parks. This type of landscape is over-protected.
(d) A landowner should be allowed to develop his own land as he wishes.

Reclaiming limestone quarries

Old limestone quarries were abandoned and left for nature to heal the visual scars on the landscape. This often took many years and even nature could not do a very good job of reclamation. The main reason is that plants cannot easily colonise vertical and very steep slopes.

Environmental landscaping can speed up the reclamation process. The vertical quarry face can be blasted to give slopes of more gentle gradient. Plants can colonise these slopes more easily. Trees and bushes can be planted to speed up natural colonisation. The quarry floor can be excavated and lined with clay to form a pond or wetland area.

Q U E S T I O N S

Core

1 Give three reasons for allowing limestone quarrying in the Peak District.

2 Match the following words with the correct letters on figure 26.5: high moor, reservoir, limestone cliff, dale, drystone walls, old limestone quarry, old mine.

3 Draw a simple sketch to show how reclamation of a derelict quarry can be speeded up.

Tokyo

Environment v. development

unit 27

... at a price

Country	per capita GNP US $
Japan	21 640
Britain	12 800
Australia	12 390
Bangladesh	170
Tanzania	160

Table 27a

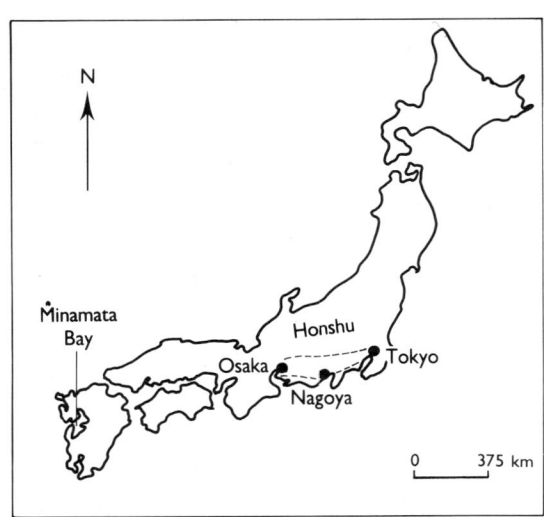

Figure 27.1   Japan

Wealth

How do you measure it? Money in the bank? Lamborghini on the drive? En-suite jacuzzi? Good friends? Nice view? Clean air?

The most common way to measure a country's wealth is to compare its per capita GNP with that of other countries. GNP stands for Gross National Product. Per capita GNP is the average amount of money generated by one person in a year. Table 27a shows some per capita GNPs in 1988.

Around the world the countries with the highest per capita GNPs are the developed industrial countries. They are seen as 'rich' countries. Countries with the lowest per capita GNPs are the less developed rural countries. They are seen as 'poor' countries. Most people believe that a country can only become rich by developing industries. 'Poor' countries are seeking to industrialize.

Industrial development causes environmental problems. That lesson has been learned the hard way by many industrial countries. Will the same lesson be learned by those countries now seeking to develop?

Japan

Japan is one of the world's richest countries. In 1930 about half of all Japan's population were farmers who lived in the countryside. By 1990 half of all Japanese people lived in the Tokyo–Nagoya–Osaka **megalopolis** in southern Honshu. One quarter of all Japanese lived in the Tokyo metropolis.

Rapid industrialisation after 1950 brought wealth, high consumption of luxury goods and the highest life expectancy to Japan . . . at a price.

► *1950s*   Mercury was dumped into Minamata Bay from a petrochemical factory. It was converted to very toxic methyl-mercury by marine organisms. This poison concentrated in the marine food chain to fish. Fish was a staple diet for local people. More than 50 people died and some children were born malformed and brain-damaged.

► *1960s*   The quality of Tokyo's air deteriorated rapidly. In July 1970 hundreds of people were taken to hospital with sore eyes and throats and breathing difficulties. The cause was photochemical smog. This is caused mainly by the action of sunlight on car exhaust fumes. Tokyo set up a smog warning system. Noise and traffic congestion were added problems.

▶ *1970s* Water in the bay became polluted with **cadmium**. This poison caused a disease called Itai-Itai which attacks bones and makes them fragile.

Action In 1972 Japan set up the Environment Agency to control pollution. It passed laws to make industries,

- pay for harmful effects of pollution

- develop pollution control technologies

Pollution controls have worked. In the 1980s less than 1% of the water samples tested failed the pollution levels and air pollution in cities was reduced to a quarter of what it had been in the 1960s. Rich countries like Japan can afford to pay for pollution control.

Pollution problems similar to those in Japan have been seen in other industrial countries.

Paper . . . at a price

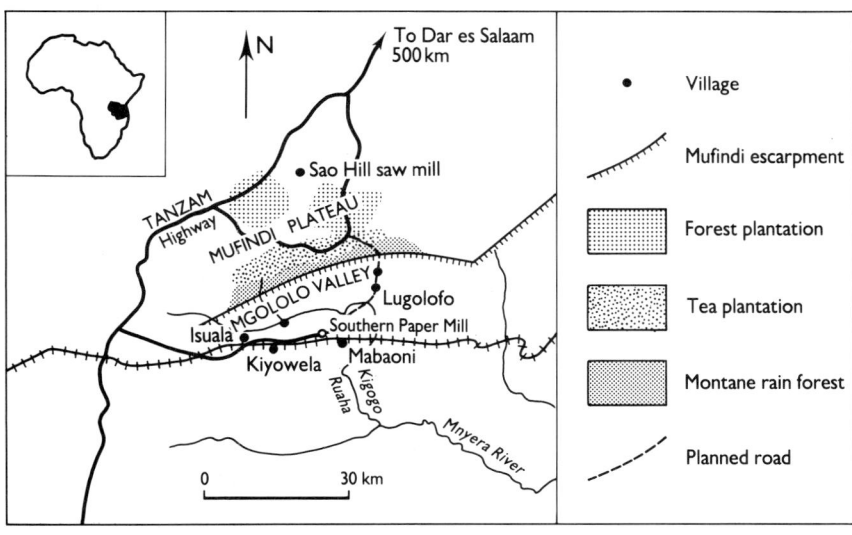

Figure 27.2 Location of Southern Paper Mill

Tanzania is a 'poor' country. It is trying to develop industrial projects. One project was to build a pulp and paper mill to produce 80% of the country's paper needs. The mill was built in an isolated rural area where local villagers depended entirely on subsistence agriculture. Problems brought by the project are:–

▶ Land – villagers lost land for the mill site.

▶ Food – local villages could not grow enough food for the mill township.

▶ Facilities – no hospital or secondary school. Poor recreational facilities.

▶ Smell – a smell of sulphur spreads from the mill.

▶ Plant life – sulphur from the mill causes local wet and dry acid deposition. There may be early signs of damage to local plants and crops, forest reserves and the tea plantation.

In spite of these problems most local people felt that the mill had brought them a higher standard of living.

Figure 27.3 shows the daily pollution created by the mill.

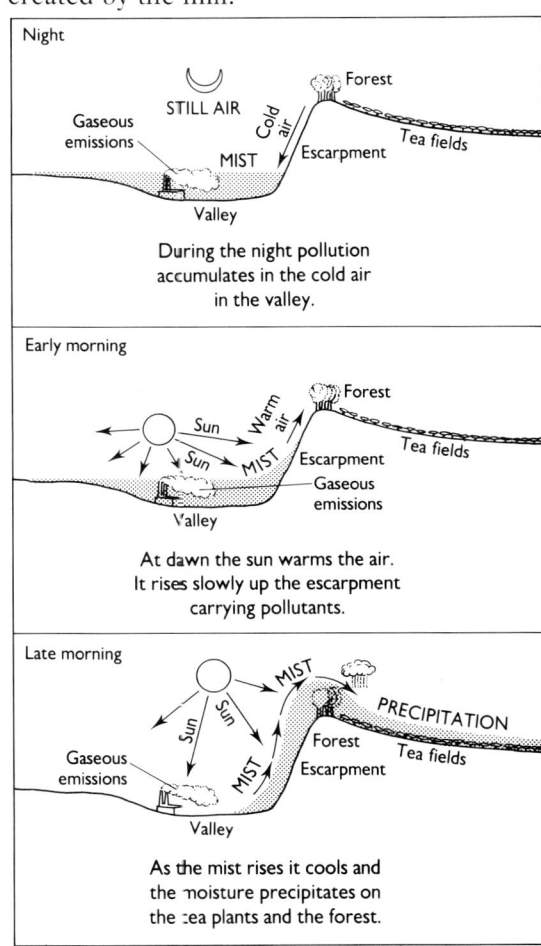

Figure 27.3 Pollution pathway

Core

1 How are 'poor' countries seeking to become 'rich'?

2 Use these words to describe problems brought to Japan by industrial growth – rich, industrial, megalopolis, Southern Honshu, Minamata disease, air pollution, Itai-Itai.

3 Which environmental problems has the Southern Paper Mill brought to local villages?

4 How can the paper mill damage the tea plantation?

5 How can industrial pollution be reduced?

The Aswan Dam

Making the most of what you've got

Figure 28.1   Quick, here's this year's supply

A harsh environment

What kind of environment does figure 28.1 show? Which two three-letter words would you use to describe its climate? Which essential resource will be in very short supply here?

If you lived in a desert you would try to make the most of any water which came your way.

Side effects

Egypt

Egypt is 96% desert. Figure 28.3 gives some climate data for the country. Total rainfall in a year in Egypt is very low. However, the world's longest river flows through the country. This river rises near the equator where the rainfall is high. The River Nile is Egypt's most important water resource – almost its only water resource. The river is the lifeblood of the country.

► 90% of Egypt's population lives on the river's flood plain.

► Almost all of Egypt's agriculture is supplied with irrigation water from the Nile.

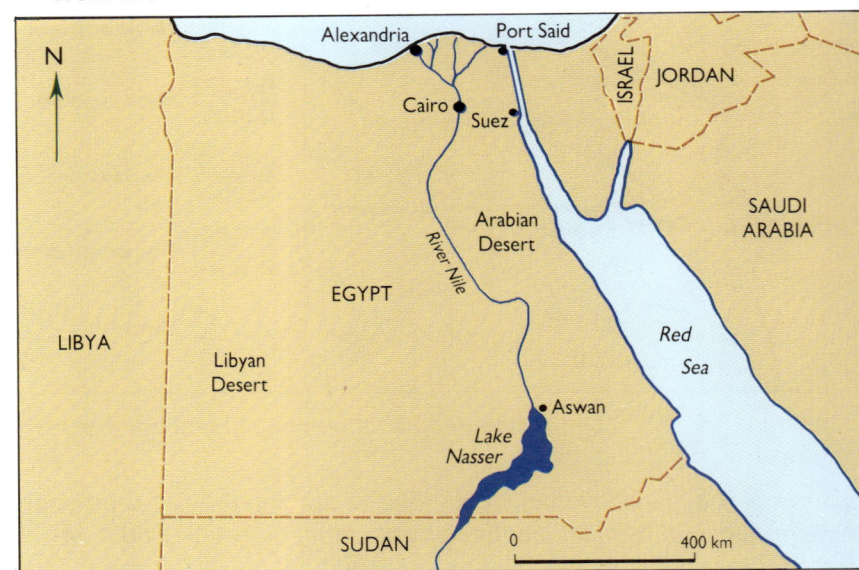
Figure 28.2   Lake Nasser

Controlling the Nile

Egyptians have always sought to control the Nile. Control of the river has passed through three technological stages. These are;

Stage 1
Before 1800 basin agriculture was practised. Banks divided the valley lands into basins. When the river flooded it deposited free fertilizer in the form of nutrient-rich silt into the basins. Only one crop could be grown each year.

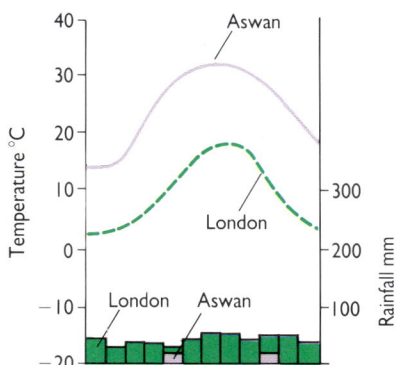

Figure 28.3   Climate data for Egypt

Stage 2

From 1850–1970 barrages, dams and canals were built. They gave better control of the Nile's flow. Some irrigation became **perennial**. Crops could now be grown all year. Maize and cotton became important. High floods were still not controlled. They caused much damage and death. The growing population also needed more food.

Stage 3

In 1970 the Aswan High Dam was built.

Aswan High Dam and Lake Nasser

The Aswan High Dam

► Was built with Russian money and technology.

► Is 111 m high and 3.7 km long.

► Was completed in 1970.

Lake Nasser

► Covers 1550 square miles.

► Is almost 300 miles long.

► Averages 14 miles in width.

► Was full in 1981.

Q U E S T I O N S

1 Copy and complete.
The River _____ is the most important water resource in _____. Its flow is controlled by the _____ _____ _____. The water is stored in Lake _____.

2 Study the climate data in figure 28.3. Write two or three sentences to explain Egypt's climate.

3 Make a copy of the map figure 28.2 in your book.

4 Copy the symbols drawn in figure 28.4. Separate them into advantages and disadvantages. Write short notes to explain each symbol.

Good intentions . . . but

Egypt needed to have better control of the River Nile to bring benefits to the country. The intention was that the Aswan High Dam would bring two particular benefits. These were to

► Increase agriculture.

► Generate electricity.

Sometimes a scheme which brings benefits also brings unintended harmful effects. The Aswan Dam brought both.

Advantages

► *Agriculture* 2 million acres of new land was brought under cultivation. ¾ million acres of basin agriculture were converted to perennial agriculture. Egypt's agricultural productivity was doubled.

► *Electricity* The dam tripled the country's electricity production.

► *Fish* Fishing is being developed on Lake Nasser. Yields may reach 10 000 tonnes per year.

► *Flooding* Flooding of the Nile's flood plain has been eliminated.

Disadvantages

► *Relocation* 90 000 people lost their homes and were relocated.

► *Fertilizer* It is no longer free. Chemical fertilizer is now used.

► *Fishing* Little nutrient now enters the Mediterranean. This has reduced the algae which in turn has seriously reduced local fish stocks by up to 95%.

► *Disease* Snails spread the disease **bilharzia**. Annual floods used to kill most of the snails. No floods means more snails means more disease.

► *Erosion* The sea is eroding the Nile delta because it is not now being built up with silt.

► *Salination* Ground water levels have dropped in the delta. Salt water has seeped into some coastal areas. Plants do not like salt in soil.

► *Siltation* Much silt now stays in Lake Nasser. It will reduce the reservoir's water-holding capacity.

Overall, most people agree that the benefits brought by this scheme outweigh the harmful effects.

Figure 28.4 Symbols

The stolen sea

Your education

Education is a partnership between the government, your school, your parents and YOU. Sometimes you should take the most responsibility for your education. Do you?

D I Y

As you know this stands for do it yourself. So, you take responsibility for this unit. Try not to ask your teacher for any help.

1 WORK in small groups of three or four.

2 STUDY the whole unit - text, diagrams, maps and sketches.

3 DISCUSS the information in the unit with your partners.

4 DECIDE what this issue is about and how you will present it in your book.

5 PRESENT it.

6 REPORT to your teacher when you have finished.

Do a good job.

Inputs = Outputs

Look at figure 29.1. It shows a drainage basin. This is the area of land drained by a river. In a natural drainage basin the amount of water going in must equal the amount of water going out. This is called the water balance. The simplest water balance has one input and three outputs. They are:

▶ Input
Water enters by precipitation – P

▶ Outputs
Water leaves by evaporation – E
transpiration – T
runoff   – R

This simple equation describes the water balance,

$$P = E + T + R$$

Figure 29.1   Inputs – outputs

In figure 29.1 you can see,

- How much water enters the basin by precipitation.
- How much leaves by evaporation and transpiration.

How much water will flow into the lake as runoff from the river? This amount of river runoff keeps the lake at a steady size. Both in depth and area.

People move into the basin. They plant crops. The climate is hot. They take 20 units of water out of the river to irrigate the crops. P = E + T + I + R How much water now runs into the lake? What will happen to the lake?

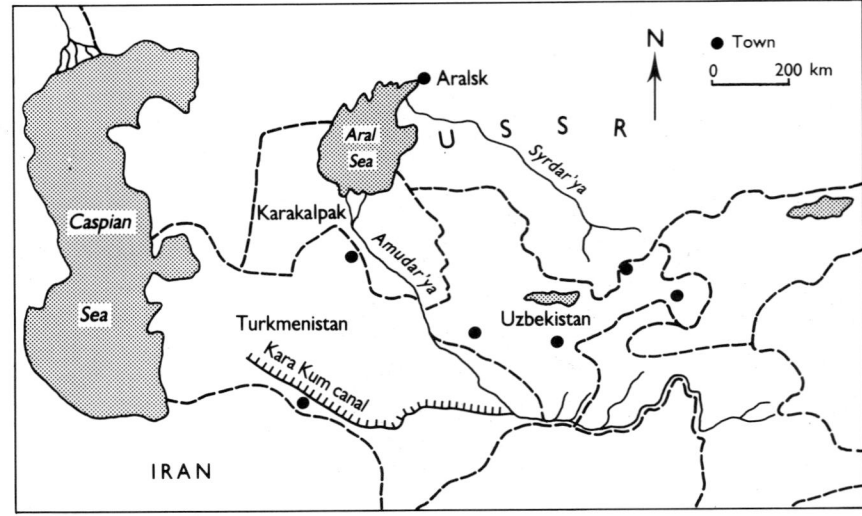

Figure 29.2   The Aral Sea

Worse than Chernobyl?

In 1989 a newspaper article and BBC television programme both described a disaster in part of the USSR. They used official Russian publications to state that –

▶ 83% of children in Aralsk had some illness.

▶ Infant mortality
 (a) in Uzbekistan was 40 per 1000.
 (b) in Turkmenistan was 58 per 1000. This is about four times the USSR average.

▶ In Karakalpak 66% of the people suffer from hepatitis, typhoid or throat cancer.

▶ The Aralsk docks are now 30 miles from the sea.

▶ Aralsk is suffering a water shortage.

▶ Fishing villages are stranded many miles inland.

▶ There are now 4 species of fish in the Aral Sea. There used to be 38 species.

▶ The Aral Sea has fallen by 14 m and has shrunk by half in the last 30 years. It is now two seas. It has lost 69% of its water.

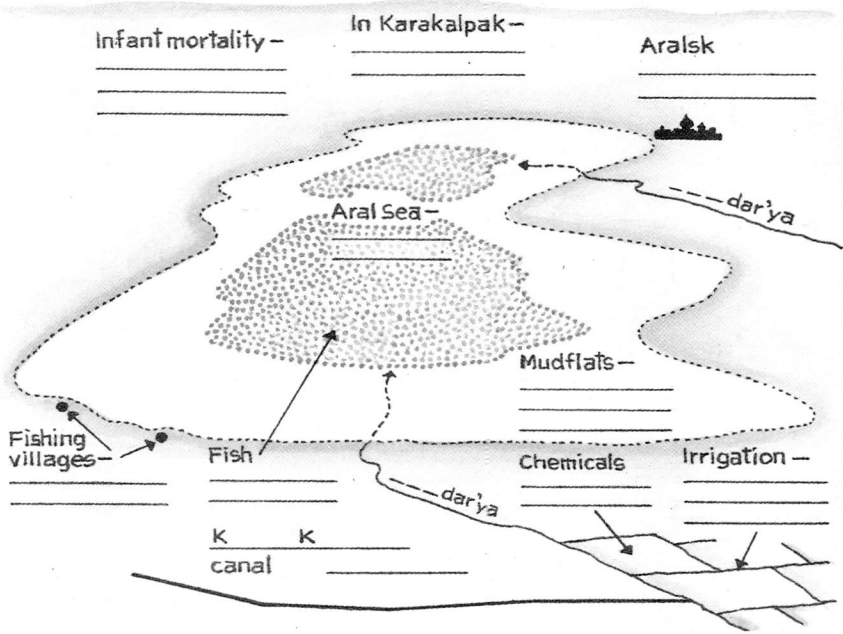

Figure 29.4   Slaves to cotton

Cotton slavery

Growing cotton and rice is said to be the cause of this disaster. Synthetic fibres are underdeveloped in the USSR. Cotton is therefore a very important crop. This region was chosen to be one of the main cotton growing areas.

Cotton and rice are very thirsty crops. To grow 1 kg of
corn needs   . . .   1400 litres of water
rice needs   . . .   4700 litres of water
cotton needs . . . 17 000 litres of water

Soviet central Asia has a dry climate, so water was drained from the region's two rivers, the Amudar'ya and the Syrdar'ya. Nearly half the water in the Amudar'ya has been channelled to the fields of Uzbekistan and Turkmenistan for irrigation. In the hot sun evaporation from open ditches is very high. Between 1974 and 1986 the Syrdar'ya did not reach the sea. Neither did the Amudar'ya between 1982 and 1989.

Massive amounts of chemical fertilizers, pesticides and **defoliants** have been applied to the crops. These chemicals have seeped into the public water supply. Some people have been poisoned. Millions of tonnes of salt and chemicals have been left on the mudflats as the Aral Sea dries up. Wind storms pick them up and dump them onto food crops and towns in the region.

OK . . . DIY. You can use figure 29.4 in your presentation if you want to.

Figure 29.3   Aral fishing fleet

Response to natural hazards

Calm after the storm

On 1 February 1953 a spring tide whipped up by severe gales caused a storm surge in the North Sea. Seawater smashed through dykes and caused disastrous flooding.

▶ 1835 people drowned.

▶ Hundreds of dykes were breached.

▶ 200 000 hectares of land was covered with salt water.

▶ Thousands of cattle drowned.

▶ Thousands of homes were destroyed.

▶ Many roads were damaged.

Figure 30.1   A breached dyke

Halt . . . friend or foe

The Dutch have a love-hate relationship with the sea. It is seen as both friend and foe. As a friend the sea has enabled south-west Netherlands to become a major centre of trade, industry and agriculture in Europe. As a foe it caused the Netherlands' worst natural disaster.

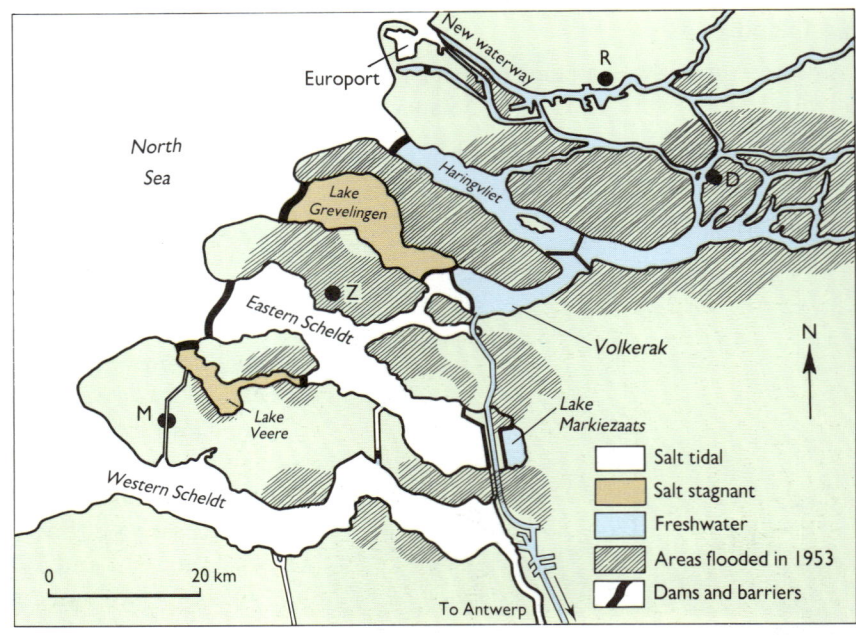

Figure 30.2   The Delta Project

The Netherlands' response – The Delta Project

The Dutch responded to this natural disaster by building the Delta Project. The aim was to close off from the sea all the tidal inlets in the delta except the New Waterway and the Western Scheldt which are the shipping routes to Rotterdam and Antwerp.

This massive plan involved building,
- five primary dams
- five secondary dams
- two new bridges.
- two storm surge barriers
- strengthening many dykes

The final part of the project was to build a dam across the Eastern Scheldt. Many people objected. They wanted to keep the tidal flow in the estuary to preserve the natural environment. A new study produced an alternative plan. The dam was rejected and a storm surge barrier was built instead.

In normal conditions the barrier remains open so that salt water can flow into the estuary. When a storm surge is expected the barrier is closed until the danger has passed.

Figure 30.3   A delta barrier

Secondary advantages
Other advantages brought by the project are,

▶ Agriculture   In the past agriculture in the delta has suffered from the seepage of salt water into fields and from water shortages in summer. Both will be prevented by freshwater estuaries.

▶ Shipping   Tidal currents have stopped. This has stabilized navigation channels, a benefit to shipping.

▶ Traffic   New road links have reduced the isolation of the islands.

Disadvantages
Creating a safe area has produced some disadvantages.

▶ Disruption to village life   Camping, water sports and second homes have increasingly caused alienation with traditional village life.

▶ Fishing ports   Some were cut off from the sea. New harbours have been built.

▶ Flora and fauna   Freshwater ecosystems are replacing saltwater ecosystems.

Q U E S T I O N S

Core

1 Study figures 30.1 and 30.2. Describe the natural hazard which caused this disaster using the following words:
- 1953 ● Netherlands ● severe gale
- storm surge ● dykes
- severe flooding ● 1800 deaths

2 Copy and complete.
The Dutch responded to this hazard by building the _____.

3 Use an atlas to name the towns M, Z, R and D on figure 30.2.
Which of these towns suffered severe flooding in the 1953 disaster?

A 1 Which of these water bodies is salt tidal, salt stagnant, freshwater?
(a) the Haringvliet   (b) Eastern Scheldt   (c) Lake Veere.

2 How does a storm surge barrier work?

B 1 Why was a barrier built on the Eastern Scheldt?
2 Can you work out from figures 30.2 and 30.4 the name of the water body marked W on the photograph?

3 Describe one secondary advantage and one disadvantage brought by the Delta Project.

Figure 30.4   Where is this?

Rising damp

Delta

- In the Greek alphabet it is the fourth letter.
- In Lancia Delta Integrale it is the car that won eight World Rally Championships in the 1980s.
- In another 50–100 years it might not be a good place to be.

Figure 31.1   Flooding in Bangladesh

Living in a delta

In geographical terms a delta is the low, flat land at the mouth of some rivers. It is formed when the river deposits sediment in the sea. The sediment builds up to form new land. Usually the river splits into many separate channels called **distributaries**.

Living in a delta can have some advantages. For example flat land makes it easier to build houses, factories, roads, railways. The soil is usually fertile so is good for crops.

But there is one serious disadvantage – flooding. Most deltas are prone to flooding. A delta will flood if the land subsides or the sea level rises.

Bangladesh

Find a map of Bangladesh in your atlas.
Bangladesh,

- Is one of the world's most densely populated countries.
- Has 105 million people living on just 144 000 km² of land.
- Has a population growth rate of 2.6%. This means the population could double in less than 30 years.
- Has 80% of its land area on the Bengal delta.
- Has half of its land less than 5 metres above sea level.
- Is one of the poorest countries in the world.

Three rivers form the Bengal delta. Every year they deposit more than 1000 million tonnes of sediment into the Bay of Bengal. Annual floods can

- drown up to 35% of the whole country.
- put some land 2 metres under water.
- damage 25–35% of the rice and jute crops.

Between 1960 and 1981 Bangladesh had 17 serious floods. The worst was in 1970. A cyclone caused very strong winds. They raised the sea level by 9 m and caused a storm surge which drowned half a million people.

Bangladesh faces the same natural hazard as the Netherlands but it is one of the poorest countries in the world, so it has been unable to respond to this hazard in the same way that the Dutch have.

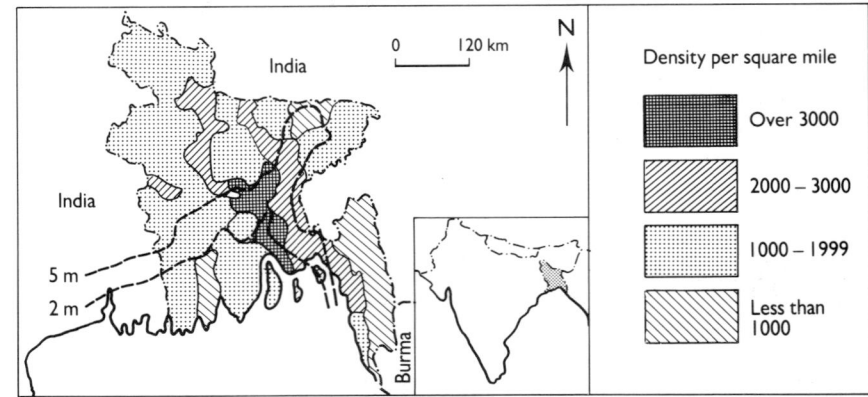

Figure 31.2   Bangladesh's population 1989

That sinking feeling

The **delta** is sinking. This is due to,

- *Local subsidence.* The causes are natural subsidence of the land and subsidence caused by people extracting water from thousands of wells.
- *Sea level rise.* A main cause could be global warming helped by the greenhouse effect.

The total rise in sea level around the coast of Bangladesh has been estimated at 1–2.5 cm per year. What could happen to Bangladesh in the future if the land continues to sink and the sea level continues to rise? We cannot travel into the future to have a look for ourselves. So what can we do?

Computer model

In 1989 scientists used a computer to study what could happen to Bangladesh in the future. They looked at three cases of sea level rise.

Best Case — A small rise in sea level.

Worst Case — A large rise in sea level.

Really Worst Case — A very large rise in sea level.

They also considered two dates, 2050 and 2100.

Table 31a shows their findings. In the Best Case in 2050 the sea level was assumed to rise a total of 13 cm. This would not cause any serious problems. Less than 1% of the land would be flooded. A few people would be displaced. Very little of the economy would be destroyed.

		2050			2100		
		Best	Worst	Really Worst	Best	Worst	Really Worst
World sea level rise	(cm)	13	79	79	28	217	217
Local subsidence	(cm)	0	65	130	0	115	230
Total sea level rise	(cm)	13	144	209	28	332	447
% loss of habitable land		<1	16	18	<1	26	34
% of population displaced		<1	13	15	<1	27	35
% of economy destroyed		<1	10	13	<1	22	31

Table 31a   Computer view for Bangladesh

However, in the Really Worst Case the sea could rise 209 cm. This would cause:

► 18% loss of land

► 15% of the population to be displaced

► 13% of the economy to be destroyed

Remember this is what could happen, not what will definitely happen.

Core

1 What is a delta?

2 Give one advantage and one disadvantage of living on a delta.

3 Use an atlas to name
   (a) the largest river which forms the Bengal delta.
   (b) the large deltas in Egypt, Burma and the USA.

4 Copy and complete.
   The Bengal delta is at risk from flooding in the future because the land is s_____ and the sea level is r_____.

5 Draw an outline map of Bangladesh   (a) in 1989   (b) with a sea level rise of 5 metres.

6 Describe what could happen to Bangladesh n 2100 if there were
   (a) a large sea level rise   (b) a very large sea level rise.

Class Study

The map in figure 31.3 shows the areas likely to be affected by a sea level rise in 2050. Discuss these points.
(a) Could this happen to Britain?
(b) Can it be prevented?
(c) What can be done if it does happen?

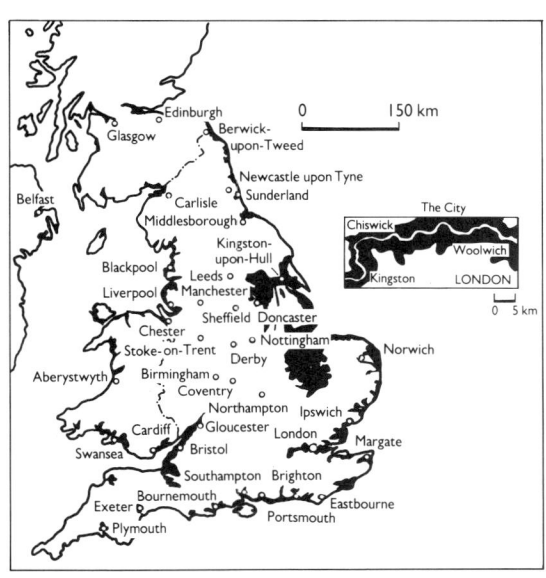

Figure 31.3   Ark's map for a greenhouse Britain

Global issues

unit 32

Help ... they're putting holes in me

Figure 32.1   Defenders of the earth

Friendly ozone?

What is it?   Three atoms of oxygen make up a molecule of ozone.

Where is it?   Most ozone is found 17–25 km above the earth in a layer in the atmosphere called the **ozone layer**.

What does it do?   The sun radiates heat and light which life on earth needs. It also emits ultra-violet rays. A small amount of ultra-violet radiation is good but too much can cause damage to people and plants. Figure 32.1 shows that the ozone layer acts like a shield and stops too much ultra-violet radiation reaching the earth.

So, ozone is friendly to us.

Ozone friendly?

Are we friendly to ozone? The answer is no. Some of the things we do are destroying the ozone layer. In September 1987 a British and American expedition found a large hole in the ozone layer over the Antarctic. Figure 32.2 shows that the hole grows and decays with the seasons. In 1987 the amount of ozone was 15% less than in 1985. In October 1987 97.5% of the ozone in the middle of the layer had been destroyed.

The main culprits
These are chemicals called **chlorofluorocarbons** or CFCs for short. When they are set free they drift up into the atmosphere and attack the ozone.

The sources
There are several sources of CFCs. Three of the main ones are drawn in figure 32.3.

Damage

What damage could be done by destroying the ozone layer?

► *People*   A main concern is that more people could develop skin cancer.

▶ *Plants*   Not much is known about the effect that too much ultra-violet radiation will have on plants. It has been suggested that even a 1% increase in ultra-violet radiation may cause some damage to plants and so affect food chains.

▶ *Atmosphere*   Ozone absorbs some of the sun's heat. This might interfere with the earth's climate.

Figure 32.3   CFC sources

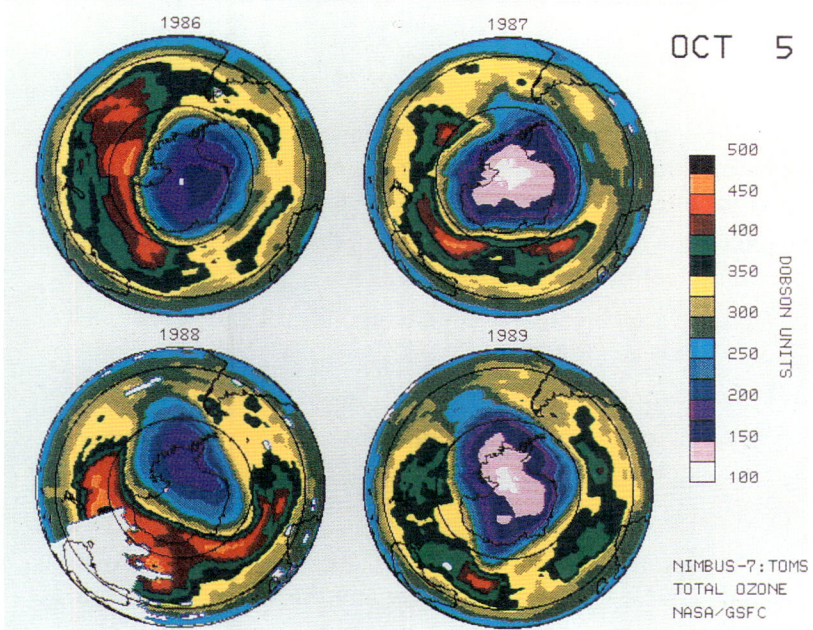

Figure 32.2   Holes in the ozone layer

What is being done?

▶ September 1987   Montreal Conference – 27 countries signed a treaty to cut CFCs to 50% of present levels by 1999. Some scientists think that this is not enough. They wanted an 85% reduction in 1990.

▶ Aerosols are being labelled 'ozone friendly' if they do not contain CFCs.

▶ 1988   Some companies and local authorities collect old fridges and remove the CFCs in a safe way.

▶ June 1990   London Ozone Conference. Developing countries argued that they do not have the technology or wealth to develop alternatives to CFCs. A deal was agreed for the rich countries to pay most of the cost of transferring technology and building new factories in developing countries. The USA was expected to pay 25% of the costs and wanted control over how the money was to be spent. Developing countries wanted to control the money themselves. China and India make up 40% of the world's population. By the beginning of 1991 neither country had signed the treaty to phase out CFCs.

ⓆⓊⒺⓈⓉⒾⓄⓃⓈ
Core

1 What is ozone and where is it?

2 What important job does it do?

3 What chemicals destroy ozone?

4 Name three sources of these chemicals.

5 Design a double page poster to explain the ozone issue. You can use the sketches in this unit to give you some ideas.

Class work

1 Look at the aerosols in your home.
   (a) Make a list of them in these groups, Furniture polish, Hair spray, Deodorant, Air freshener, Paint spray, Other.
   (b) Note which are ozone friendly.
   (c) Do a class survey of the results. Work out,

   ● which group contains most aerosols.

   ● what percentage of all aerosols are 'ozone friendly'.

2 Cutting down on the use of aerosols would be an alternative way of helping to reduce the threat to the ozone layer. Make a list of the aerosols you and your family use and try to think of alternatives.

Accidents will happen

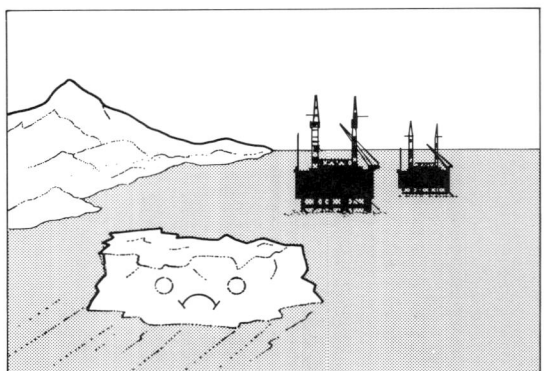

Figure 33.1   Ah! No brakes, no steering

Figure 33.2   Emperor penguin and chick

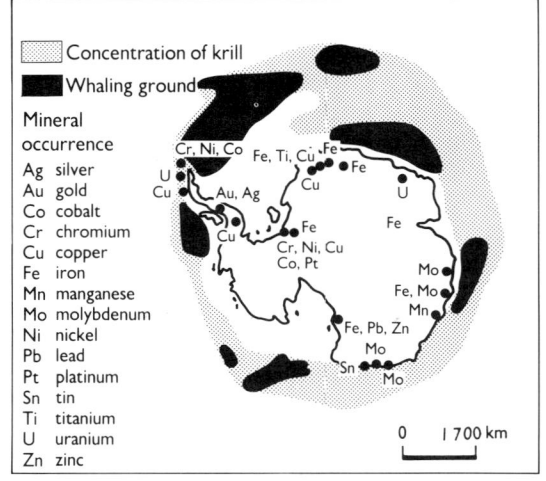

Concentration of krill

Whaling ground

Mineral
occurrence
Ag silver
Au gold
Co cobalt
Cr chromium
Cu copper
Fe iron
Mn manganese
Mo molybdenum
Ni nickel
Pb lead
Pt platinum
Sn tin
Ti titanium
U   uranium
Zn zinc

0   1 700 km

Figure 33.3
Commercial exploitation – the future

Will they or won't they?

Look at figure 33.1. What could happen? It would be nice to be able to prevent all accidents but this is not possible – accidents will happen.

Antarctica. The last resource frontier

Some people are concerned that accidents caused by humans that have harmed the environment in other continents could happen in Antarctica unless the right decisions are made now.

The past
Unlike other continents Antarctica was never developed because,

▶ It has a severe climate. Average temperatures can reach $-40°C$. Winds can reach 200 mph.

▶ It is very mountainous.

▶ 98% of the continent is covered in ice.

▶ It is isolated from the other continents.

▶ It did not seem to have any resources.

Even so, seven countries claimed Antarctica: Argentina, Australia, Chile, France, New Zealand, Norway and the United Kingdom.

The present
The rules for the continent are set out in the Antarctic Treaty which came into force in 1961. The main rules are:

▶ Antarctica should be forever used for peaceful purposes.

▶ Antarctica is not owned by anybody.

▶ Any country can set up a base camp to carry out scientific research anywhere in Antarctica.

These are not real rules. They are more a 'gentlemen's agreement'. There are no international laws or penalties to stop any country carrying out any activity it wishes in Antarctica. The Treaty did not set out any proper guidelines for wildlife conservation or mineral development.

Antarctica does not have a large number of species like the tropical rainforests. However, it does have large populations of seabirds, seals and whales. All wildlife lives in the sea surrounding Antarctica or on the narrow coastal fringe.

The Antarctic ecosystem is very fragile. Krill is the most important part of all foodchains. If the krill population was reduced through overfishing or pollution this could be serious for all wildlife.

The future
The old view that Antarctica has no commercial value has changed. Geologists think the continent could be rich in resources hidden under the ice. Resources thought to be there include coal, oil, gas, gold, iron and other minerals, and the most precious mineral of all,

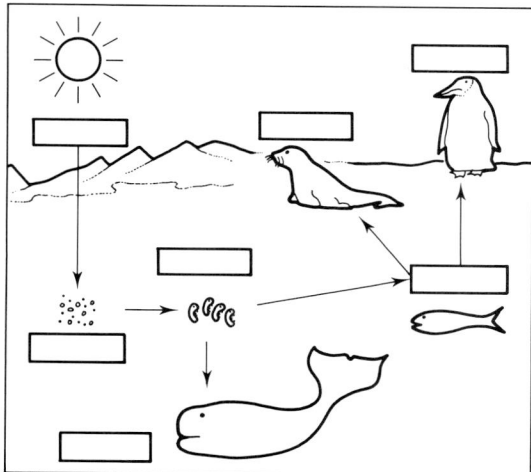

Figure 33.4   Antarctic food chain

platinum. With modern technology the severe climate and ice are no longer such great barriers to development.

There is a growing feeling that some scientific bases are not carrying out pure research but simply looking for ways to develop mineral resources.

The views

There are two main views about Antarctica's future. They are:

▶ Controlled Exploration View.

Some governments and large international companies argue that exploitation cannot be stopped. They say the best way to protect Antarctica's environment is to allow exploitation but have strict environmental controls.

Britain and the USA support this view.

▶ World Park View.

Some governments and environmental groups like Greenpeace want complete protection for Antarctica. They want it to be made a World Park. They would permanently ban:

● All commercial mineral exploitation.
● All nuclear activity.
● All nuclear and toxic waste disposal.
● All military activity.

Australia, France and Italy support this view.

Recent events

In 1988 an Argentine supply ship sank in Antarctic waters. A twelve mile oil slick spread near an important breeding colony for penguins and seals.

In December 1989 the Japanese whaling fleet started to kill 300 minke whales despite international protests – there is a world ban on whaling but a loophole in the law allows killing for scientific research.

Core

I Why was Antarctica not developed in the past?
2 Why is Antarctica an issue now?
3 Explain   (a) the Controlled Development Viewpoint
          (b) the World Park Viewpoint.
4 Copy and complete figure 33.4 by correctly labelling the organisms.

Role Play

Divide the class into five groups. Discuss the views of your group. Decide who in the group will present each point. The teacher will chair the meeting.

I World Park supporters. You:
  ▶ are concerned that mineral exploitation will harm Antarctica.
  ▶ want Antarctica left as a wilderness.
  ▶ want Antarctica made into a World Park.
  ▶ think international laws must be passed.
  ▶ want pressure put on all countries to sign a treaty agreeing to these laws.

2 Controlled Development supporters. You:
  ▶ believe exploitation cannot be stopped.
  ▶ argue that you cannot pass international laws stopping development.
  ▶ want to know how such laws could be enforced if they were passed.
  ▶ want to know what penalties would be handed out to countries who broke the laws.
  ▶ suggest the world is running out of many resources. Antarctica could have vast resources. These are needed for development to give people a better standard of living.
  ▶ suggest that people are more important than penguins, seals and whales.

3 Greenpeace supporters. You:
  ▶ want wildlife and wilderness protected.
  ▶ see mineral development as a serious threat to this.
  ▶ are concerned that countries with nuclear and toxic waste disposal problems might see Antarctica as a good dumping ground.
  ▶ want Antarctica kept in pristine condition for atmospheric research.

4 New Zealand ecologists. You:
  ▶ explain the fragile Antarctic ecosystem.
  ▶ say that a major oil spill could cause serious damage and be extremely difficult, perhaps impossible to clean up.
  ▶ give examples of any recent events which have threatened the Antarctic ecosystem.

5 Teacher and rest of class. You judge this issue.
  ▶ Listen to all the points of view.
  ▶ Discuss them.
  ▶ Take a vote on what you think should be done.

Road to nowhere for the polar bear ... or just a hoax?

Can you guess what might be the link between figures 34.1 and 34.2? If not read on.

Figure 34.2   More waterfront than they bargained for

Figure 34.1   Just a mint advert?

Greenhouse gases

Several greenhouse gases occur naturally in the atmosphere. Water vapour is the main one: It enters the atmosphere by evaporation. Some of our activities are adding to the natural concentration of these gases in the atmosphere. Figure 34.3 shows some of these greenhouse gases.

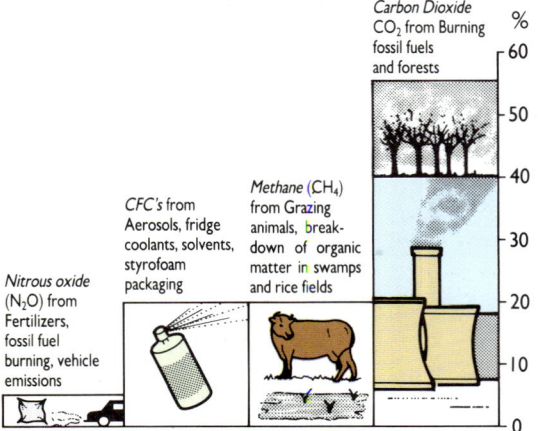

Figure 34.3   Greenhouse gases

Living in a greenhouse

Go into a greenhouse on a cold but very sunny day and you soon feel much warmer. This is because the glass lets in sunlight to heat up the inside but stops some of the heat from escaping. Greenhouse gases in the atmosphere act in much the same way as the glass. They let in the sun's short-wave light radiation to heat up the earth's surface and atmosphere but stop some of the earth's long-wave heat radiation from escaping back to space. It is a good job they do. Without them earth's temperature could be a chilling $-18°C$ instead of the present average of 15°C.

Some scientists think that if the concentration of these greenhouse gases in the atmosphere increases, more heat would be trapped and earth's temperature would rise.

Getting hotter – any evidence?

There is much controversy among scientists about the evidence for global warming. Three main areas of argument are;

1   Temperature rise.

▶ Greenhouse scientists say that earth's temperature has risen 0.5–0.9°C since 1800.

▶ Other scientists say there has been no overall change this century and that earth's temperature has fallen in the last 30 years.

2   Sea level rise.

▶ Greenhouse scientists claim a 6–8 cm rise in world sea level since 1900.

▶ Others argue that there is no clear evidence of this.

3   Carbon Dioxide.

▶ Greenhouse scientists argue that the concentration of CO_2 in the atmosphere has risen from 280–330 ppm since 1800. They argue that more CO_2 must lead to global warming.

▶ Other scientists claim that past climatic evidence suggests that a warmer world produces more CO_2 in the atmosphere, **not** that more CO_2 in the atmosphere produces a warmer world.

Three questions

1 Do you think there is clear-cut evidence that people's activities are/are not causing global warming?

2 Could any climate changes be natural? After all, the earth's climate has had several warm and cold spells in the past.

3 If the world was heading into a cold period like the Ice Ages of the past would global warming caused by human activity be a good thing?

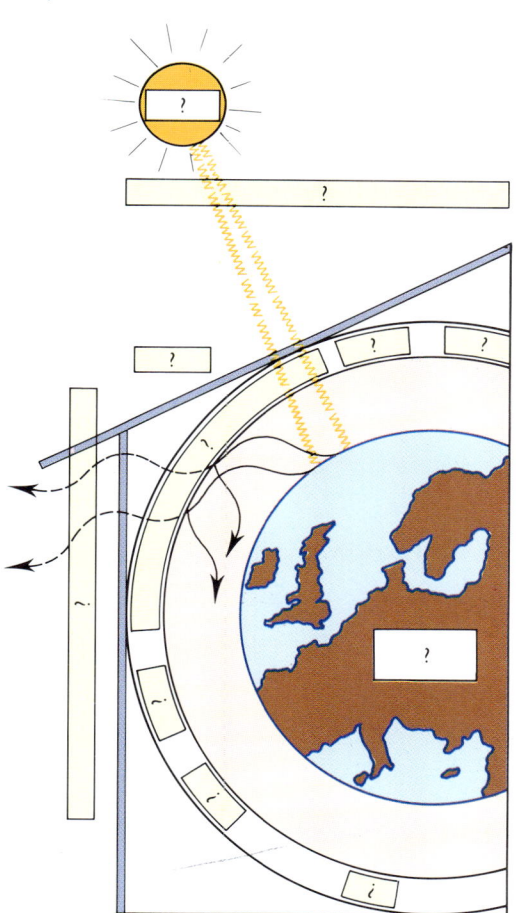

Figure 34.4   Living in a greenhouse

Q U E S T I O N S

1 Draw and correctly label figure 34.4. Use these labels.
- Sun • Earth • CO_2 • CFCs • CH_4
- N_2O • 6000°C • 15°C • greenhouse gases • short-wave light radiation • long-wave heat radiation

2 Name the three physical effects that could happen if global temperatures rise.

Problems of Measurement

1 Temperature rise.   Many world temperature measurements are taken in cities. This century cities have grown and city temperatures have risen. How much of this rise could be due to global warming and how much due to the city 'heat island', much of which is produced when the sun's heat is stored in the city's concrete?

2 Sea level rise.   In some areas of the world land is rising. In other areas land is sinking. It is therefore very difficult to accurately measure changes in sea level.

3 Computer models.   The computer models used to model the extremely complicated workings of the atmosphere are as yet too simple. They cannot take into account all the factors which together determine earth's climate.

A warmer world?

No one knows for certain what will happen to earth's climate in the future. However, we should try to plan for all possibilities. What could happen to our planet if the greenhouse scientists have got it right? By 2050 they have predicted a global temperature rise of 2–5°C.

► Water in the oceans could expand. Sea levels could rise by 1 m in the next 50 years.

► Ice caps could start to melt. Sea levels could rise even more.

► World climates could change. Some areas could become warmer, some colder, some wetter, some drier.

Class discussion

Global warming – has it started?
is it due to man's activities?
is it a bad thing?
Make brief notes of any points you think are important.

Research

1 In pairs.
(a) Design a simple questionnaire to find out what people know and how concerned they are about the greenhouse effect. Here are some ideas to get you started.
You could ask them if they can name any greenhouse gases.
Do they know where these gases come from?
Can they describe any effects of global warming?
Are they worried about global warming?
(b) Use your questionnaire to carry out a local survey.
(c) Record the results.
(d) In class discuss your findings.

2 How much do people know and care about this very important global issue?

Possibilities or probabilities?

This page shows some possibilities for a 'Greenhouse World'.

1 On a double page in your book draw a map to show possible world wide impacts of global warming.

2 In small groups,
   (a) Obtain an OS map of a low-lying coastal area. For example, the Wash, Solway Firth or Somerset Levels.
   (b) Imagine a rise in sea level of 5–10 metres.
   (c) Discuss what could happen to any town, villages, farmland, valuable resources.

3 Draw up a list of suggestions of what can be done to help reduce global warming..

Arctic Global warming could melt the polar ice. A warmer ocean would produce less plankton. Plankton is the base of the ocean food chain. There would be less krill, less fish and fewer seals. No ice bridges and few seals would make life difficult for the polar bear.

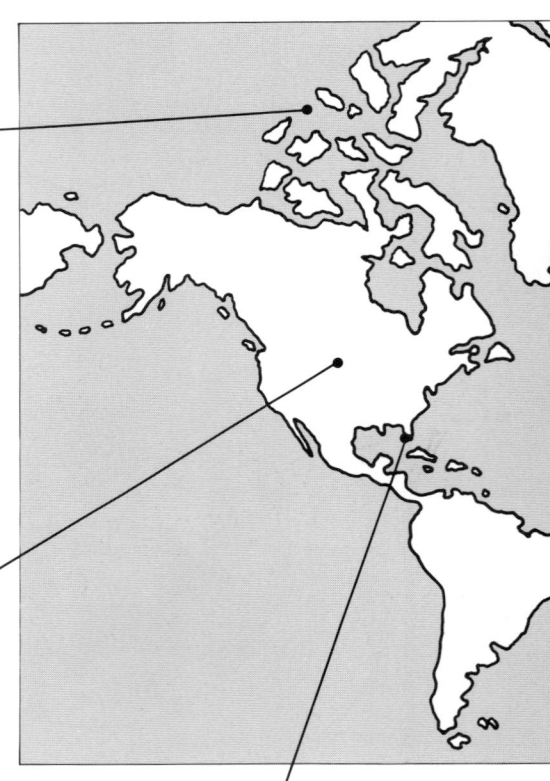

Drought Would global warming produce more droughts in America? How would this affect wheat production in the Prairies?

The Everglades Will this low-lying alligator habitat be swamped? Alligator eggs incubated at less than 30°C become females. Eggs incubated at about 35°C become males. What might happen in a warmer world?

Holland Two thirds of this country is below sea level. Barrages and dykes have been built to stop the sea from invading. Can they keep out rising seas?

Coniferous forests In a warmer climate they could move northwards.

Bangladesh Half of this country is less than 5 m above sea level. Unit 31 showed how a rising sea level could affect Bangladesh.

Environmental refugees In the past people have left their countries for political and economic reasons. In the future will they have to leave for environmental reasons?

Coral islands Coral grows very slowly only near the surface of the sea. Can it keep up with a rising sea level?

Pests Insects and pests could be winners in a warmer world. Will locusts move into Europe from Africa?

Sustainable development

Figure 36.1   Sustainable relationships?

Earth's population

Table 36a shows the growth of the human population on the planet in the last four hundred years. The populations of different countries are growing at different rates. Populations in some countries are increasing very rapidly. Other countries have almost reached zero growth.
Table 36b shows different growth rates around the world.

Year	Millions
1700	500
1800	900
1900	1600
2000	6000 (estimated)

Table 36a   Human population growth

Bacteria, lemmings and wolves

Bacteria can do this every twenty minutes: 1, 2, 4, 8, 16, 32, ?, ?, No they cannot count. Yes they can double their population. Some bacteria were put in a test-tube with a food resource. At first they increased rapidly. Then they all died out. Even before the food ran out.

Norway's lemming population increases rapidly every few years. Most of the food is eaten. Facing starvation the lemmings migrate. Many migrate into the sea. Unfortunately, lemmings cannot swim very far.

Wolves living on an island in Canada prey on a herd of caribou. When the caribou population is very low the wolves do not breed. Both animals have lived on the island for many years.

Questions

1. Which of these three cases has a sustainable relationship with its environment?

2. Can you suggest why all the bacteria died?

Spaceship Earth

Earth has been called a spaceship which cannot return to base. On board is a supply of resources. Some are finite and will eventually run out. Some are renewable and must not be polluted so they will stay renewable. These resources have a **carrying capacity**. This means they can support so many people living at a certain level of consumption for a certain time.

More people and a higher consumption simply means the resources are used up more quickly. More consumption also produces more pollution. Carrying capacity is about,
● population ● consumption ● resources ● pollution.

Figure 36.2 shows some of earth's resources and the rate at which they are being used up.

By the way, the bacteria produced wastes which polluted their environment and killed them off.

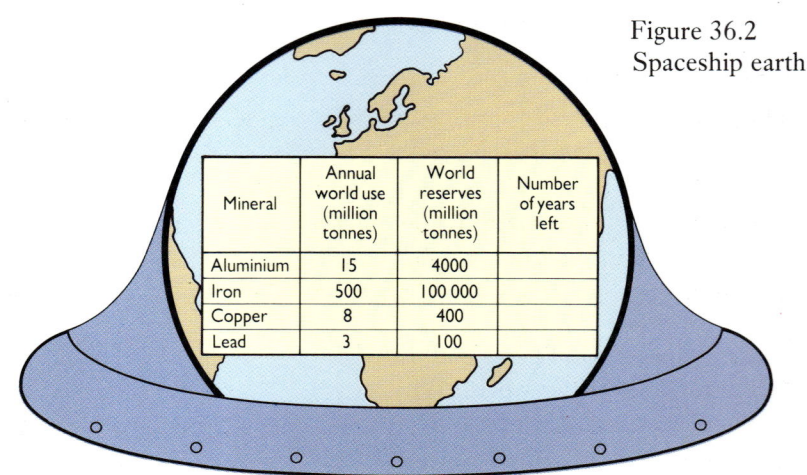

Figure 36.2 Spaceship earth

Mineral	Annual world use (million tonnes)	World reserves (million tonnes)	Number of years left
Aluminium	15	4000	
Iron	500	100 000	
Copper	8	400	
Lead	3	100	

Country	Population (millions) 1988	(est) 2000	Growth rate % per year
USA	246	263	1.0
Brazil	144	178	2.2
UK	57	59	0.2
Pakistan	106	146	3.1
Nigeria	110	164	3.4
USSR	286	308	0.9
China	1084	1274	1.2
India	814	996	2.1
Indonesia	175	212	2.1
Australia	17	18	1.4

Table 36b   Population growth rates

Development

Development aims to improve what is called,

▶ The quality of life.   ▶ The standard of living.

Are these the same thing?

Everybody on earth has a right to live in an environment which provides,

● Enough good food.   ● Clean air and water.

● Employment.   ● Good housing and sanitation.

● Good health and recreational facilities.

Some people would also want a car, a big house, TV, video, holidays, new clothes and furniture and many other things. Other people might only want friendship, trust, respect, religion. For still others it might be a mixture of both of these views.

Sustainable development

Sustainable development means developing in a way that protects the environment both for people and for other creatures which share earth with us. It should,

▶ Use finite resources wisely to make them last as long as possible.

▶ Protect renewable resources from pollution so that they stay renewable.

▶ Replace finite with renewable resources where possible.

▶ Share resources with others.

▶ Avoid pollution wherever possible.

When you are 20

Think of three things you would like to have when you are twenty. How many in the class included a car? What role does a car play in spaceship earth?

The car consumes more resources than any other industry. In some countries cars can consume:

● 20% of all steel   ● 10% of all aluminium

● 35% of all zinc   ● 50% of all lead

● 60% of all natural rubber   ● 50% of all oil

Think how much of the world's resources will be used if every family on earth owned a car.

Q U E S T I O N S

Core

1 What do you think is meant by the term 'sustainable development'?

2 Copy figure 36.2 into your book and fill in the last column.

3 Use table 36a to draw a histogram of world population growth. Project the histogram to 2100.

Small group discussion

(a) Could what happened to the bacteria happen to people or earth?

(b) What do you think is needed for a good quality of life?

Class discussion

(a) How important is the motor car?

(b) Suggest ways in which the number of cars can be reduced.

A Study table 36b.

1 What was Brazil's population in 1988?

2 How many more Brazilians might there be in 2000?

3 Which country,
   (a) had the largest population in 1988?
   (b) has the highest growth rate?

B Study table 36b.

1 Which country has,
   (a) the highest growth rate.
   (b) the lowest growth rate.

2 Calculate the percentage increase in Brazil's population between 1988 and 2000.

3 Calculate Nigeria's likely population in 2010.

Stewardship

Figure 37.1   Man or God?

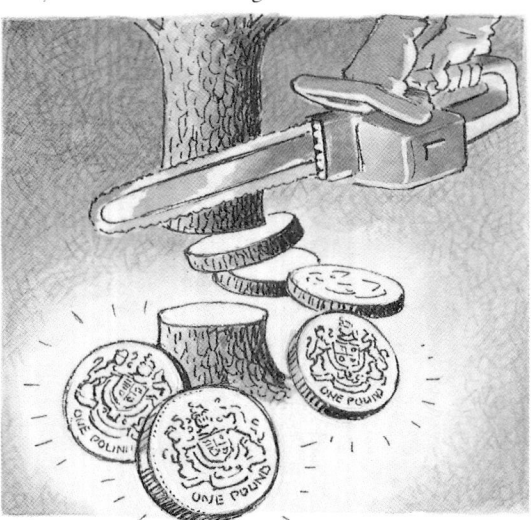

Figure 37.2   Nature has a price

Figure 37.3   Animal rights?

Most intelligent – most powerful

Put the following creatures in order of
(a) Intelligence   (b) Strength
Polar bear, human, dog.

A human would never win a fight with a polar bear but in one way man is much more powerful. What can a human do that none of the others can do? You are right if you said that a human can use intelligence to,

* make things
* adapt to almost any environment
* change most environments

People are the most powerful creatures on earth. Does this mean that:

► We have a right to use all other creatures on this planet for our own benefit?

Or

► Should we recognise that we share this planet with millions of other creatures who also have rights which we should protect?

Stewardship

Stewardship adopts the second of these views. Many people think we have a moral responsibility to manage the earth in a way that protects all creatures living on the planet.

Viewpoints – interpretations

In small groups,

1 Study the four sketches, figures 37.1, 37.2, 37.3 and 37.4.

Figure 37.4   The techno – fix

2 Discuss what you think each sketch means.

3 Your group has been invited to produce a programme for either,
   (a) A local radio station.   (c) A local TV station.
   (b) A primary school.   The title is **Stewardship**.

4 Write a script to explain Stewardship to your chosen audience. You may wish to use sketches, diagrams and photographs.

5 If you can, either,   ● Record your script on tape.

   ● Make a video.

   ● Present your programme to primary children.

Test yourself

Chapter 1–3

True or false?

1 Coal is a renewable source of energy.
2 Opencast mines coal from near the surface.
3 Visual intrusion is a main impact of opencast.
4 The *Exxon Valdez* was a container ship.
5 Drax burns oil.
6 CO_2 is the main gas causing acid rain.
7 Sweden is a victim of acid rain.
8 Chernobyl was the site of a tanker accident.
9 Acid rain can damage trees and fish.
10 A radiation pathway is grass → cow → milk → people.

Take a letter

▶ Read the question.
▶ Write down the answer.
▶ Take out of the answer the letter indicated.
▶ Use the letters you take out to unscramble the word.

Questions	No. of letters in the answer	Take out letter no.
1. A liquid fossil fuel.	3	2
2. A new coalfield in North Yorkshire.	5	2
3. The site of the world's worst nuclear accident.	9	5
4. Dangerous waste from a nuclear power station.	9	4
5. A method of mining coal.	8	8
6. Coal is this for Drax.	4	1

The answer is _?_ _?_ _?_ _?_ _?_ _?_ .

Choices

Which choice in the brackets is correct?

1 Deep coal mines may cause (subsidence, incidence, submergence).
2 Opencast mines may cause (oil, radiation, noise) pollution.
3 Most oil at sea comes from (rivers, tanker accidents, natural sources).
4 Chernobyl is in (USA, USSR, Italy).
5 In France 30 000 people a year die from (radiation, lightning strike, smoking).
6 Acid rain is produced by (NO_x, CO_2, CH_4).
7 Selby coal is clean so there is no need for a (refuse, soil, waste) tip.
8 (Overburden, topsoil, coal) acts as a screen around an opencast site.
9 Flue-gas cleaning equipment would reduce (radiation, noise, acid rain).
10 Acid rain and nuclear radiation are examples of (transylvanian, translucent, transboundary) pollution.

Chapter 4

True or false

1 Wind is a renewable form of energy.
2 Aerogenerators convert power in the wind into electricity.
3 A windfarm produces SO_2.
4 The energy content of 2.5 dustbins = one bag of coal.
5 The Severn estuary is a good site for wave power.
6 Geothermal energy is produced by gravity.
7 Iceland uses geothermal heat for district heating.
8 A windfarm might interfere with TV reception.
9 A tidal barrage harnesses solar power.
10 Tidal power is renewable.

Take a letter

Questions	No. of letters in the answer	Take out letter no.
1. Heat from underground.	10	2
2. Good site for a tidal barrage.	7	5
3. A group of windmills.	8	6
4. A country which is developing windpower.	7	3
5. Most of our rubbish is put in these sites.	8	1
6. A country with huge geothermal resources.	7	3
7. Wildlife which could be affected by the Severn barrage.	5	1
8. What do wind turbines generate.	11	3
9. Opposite of finite.	9	5

The answer is ?   ?   ?   ?   𝒆   𝒲

Choices

1 An aerogenerator is driven by (petrol, sunlight, wind).

2 Rubbish can be turned into (RDF, rifle, cattle) pellets.

3 In a barrage the (high, neap, ebb) tide flow would generate electricity.

4 (Magma, Magna Carta, Magnetic field) produces geothermal heat.

5 Windfarms cannot work (at night, in calm conditions, in strong sunshine).

Chapter 5 and 6

True or false

1 The two sources of water are surface and underground.

2 Water is stored in a dam.

3 A class 4 river would be good for fish.

4 The River Rhine flows mostly through Italy.

5 The River Rhine used to be called Europe's 'open sewer'.

6 Sewage can pollute water at bathing beaches.

7 Wood is an important fuel in many African countries.

8 Tropical rainforests have few species of plants and animals.

9 Brazil has large areas of tropical rainforests.

10 Cutting down tropical rainforests would add to the greenhouse effect.

Take a letter

Questions	No. of letters in the answer	Take out letter no.
1. Tropical rainforests are cut down for this resource.	4	1
2. A river which pollutes the North Sea.	5	5
3. An African country which uses wood as a main fuel.	5	2
4. Livestock ranched on cleared rainforest.	6	2
5. A fertilizer washed from farmland into rivers.	8	6
6. Finite fuel from the North Sea. Not oil.	3	3

The answer is ?   ?   ?   ?   ?   ?

Choices

1 Hot water enters rivers from (power stations, oil refineries, swimming baths).

2 When bacteria decompose sewage (oxygen, hydrogen, nitrogen) is used up.

3 Rainforests grow near the (North Pole, South Pole, equator).

4 Rainforests are (increasing, decreasing, remaining the same) in area.

5 Most of Britain's rivers are (good, poor, bad) quality.

6 The body which polices our rivers is called (River Pollution Board, Anti-Pollution Committee, National Rivers Authority).

7 Burning toxic waste at sea is called (incineration, incarceration, ignition).

8 Tropical rainforests hold a valuable (pool table, gene pool, blood bank).

9 The Southern Paper Mill is in (South Africa, Turkey, Tanzania).

10 Pollution from this paper mill might damage a local (tea, coffee, cotton) plantation.

Chapters 7–12

True or false

1 Soil is eroded by sunlight.

2 Pesticides are used to kill insects and weeds.

3 The Aswan Dam is in India.

4 The Aral Sea has increased in size this century.

5 The Netherlands' Delta project builds space satellites.

6 Much of Bangladesh is on a delta.

7 A hole is developing in the earth's ozone layer.

8 Antarctica may have rich mineral resources.

9 SO_2 is the main greenhouse gas.

10 The greenhouse effect could produce rising sea levels.

Take a letter

Questions	No. of letters in the answer	Take out letter no.
1. Means development which can be maintained.	11	11
2. Main problem of living in a delta.	8	3
3. Greenhouse effect could melt it.	3	1
4. Huge lake in Egypt.	6	6
5. Means looking after the earth.	11	1
6. Protects us from ultra-violet radiation.	5	5
7. Country that could be in trouble with global warming.	10	6
8. The last resource frontier.	10	5
9. Has to be used in Egypt to replace 'free' silt.	10	1
10. Crop blamed for the Aral problem.	6	3

The answer is ?   ?   ?   ?   ?   ?   ?   ?   ?

Choices

1 (BBC, TSB, DDT) is a pesticide.

2 Using another organism to control a pest is called (organic, sociological, biological) control.

3 (ATCs, CFCs, UVCs) destroy ozone.

4 Greenpeace supports a (World Zoo, World Park, Controlled Exploration) view for Antarctica.

5 About (5%, 10%, 33%) of the world's food is damaged by pests.

6 Bilharzia is a water-borne disease carried by (snails, frogs, crocodiles).

7 The River Nile has been dammed at (Cairo, Alexandria, Aswan).

8 The Aral Sea is in (Egypt, USSR, Japan).

9 Soil in rainforests can be eroded by heavy (rainfall, elephants, metal).

10 The Dutch built the Delta Project to stop flooding by the (Irish, Aral, North) Sea.

Index

Numbers refer to units not pages.

In figure 15.1 the top photo shows a geothermal bread oven.